Watershed Hydrology, Management and Modeling

Editors

Abrar Yousuf

Manmohanjit Singh

Punjab Agricultural University
Regional Research Station
Ballowal Saunkhri, SBS Nagar
Punjab, India

CRC Press
Taylor & Francis Group
Boca Raton   London   New York

CRC Press is an imprint of the
Taylor & Francis Group, an **informa** business

A SCIENCE PUBLISHERS BOOK

CRC Press
Taylor & Francis Group
6000 Broken Sound Parkway NW, Suite 300
Boca Raton, FL 33487-2742

First issued in paperback 2021

© 2020 by Taylor & Francis Group, LLC
CRC Press is an imprint of Taylor & Francis Group, an Informa business

Library of Congress Cataloging-in-Publication Data

Names: Yousuf, Abrar, 1989- editor. | Singh, Manmohanjit, 1971- editor.
Title: Watershed hydrology, management and modeling / editors: Abrar
   Yousuf, Manmohanjit Singh.
Description: Boca Raton, FL : CRC Press, Taylor & Francis Group, [2020] |
   Includes bibliographical references and index. | Summary: "The book will
   provide the comprehensive insight of the watersheds and modeling of the
   hydrological processes in the watersheds. This book will cover the
   detailed concepts of watershed hydrology and watershed management. The
   basic concepts of soil erosion and its types, measurement and estimation
   of runoff and soil loss from the small and large watersheds will be
   discussed. Recent advances in the watershed management like application
   of remote sensing and GIS and hydrological models will be included in
   the book. The insight to the various important hydrological models will
   also be given in the book. The book will be a guide for professional and
   competitive examinations to Under Graduate students of Agriculture and
   Agricultural Engineering and Master students of Soil Science/ Soil and
   Water Engineering/Agricultural Physics/ Hydrology/Watershed
   Management"-- Provided by publisher.
Identifiers: LCCN 2019032196 | ISBN 9781138365643 (hardcover)
Subjects: LCSH: Watershed hydrology. | Watershed management. | Soil
   erosion.
Classification: LCC GB980 .W368 2020 | DDC 551.48--dc23
LC record available at https://lccn.loc.gov/2019032196

Visit the Taylor & Francis Web site at
http://www.taylorandfrancis.com

and the CRC Press Web site at
http://www.crcpress.com

Preface

The concept of a watershed as a hydrologic unit advanced in the early twentieth century has been widely adopted as the management unit. Watersheds are hydrologic units that are considered efficient and appropriate for the assessment of available resources and subsequent planning and implementation of various development programmes. Most development programmes have worked along political and administrative units. Districts, blocks and villages formed the unit for planning and implementation of development programmes by the Government as well as Non-Government agencies. However, the environment does not recognize these boundaries therefore; any environment regeneration programme has to take into account environmental boundaries for effective success. The best environmental unit for planning is the watershed.

There is a need to understand physical process of erosion in relation of topography, land use and management to come up with best management practices. Planned land use and conservation measures to optimize the use of land and water resources help in increasing sustainable agricultural production. However, to achieve this, quantification of runoff and soil loss from the watersheds is must. Since it is very often impractical or impossible to directly measure soil loss on every piece of land, and the reliable estimates of the various hydrological parameters including runoff and soil loss for remote and inaccessible areas are tedious and time consuming by conventional methods. Therefore it is desirable that some suitable methods and techniques are used/evolved for quantifying the hydrological parameters from all parts of the watersheds. Use of mathematical hydrological models to quantify runoff and soil loss for designing and evaluating alternate land use and best management practices in a watershed is one of the most viable options. The integrated outcome of the hydrological models along with remote sensing and GIS can be helpful to the decision makers to evaluate the best management practices and design the necessary soil and water conservation structures to reduce the soil erosion.

The present book has been divided into 11 chapters. The book covers both the basic and applied parts in relation to watershed hydrology. The basics of soil erosion, measurement of soil erosion, runoff and rainwater harvesting and basic information about watershed hydrology and management is given, in order to understand the basic processes in relation to watershed management. The applied part includes land evaluation, modeling soil erosion, case studies on use of various models like Erosion 3D model and SWAT model and futuristic approach to watershed management. The comprehensive insight of the watersheds and modeling of the hydrological processes in the watersheds and the detailed concepts of watershed hydrology and watershed management, recent advances in the watershed management such as the application of remote sensing and GIS and hydrological models have been included in the book. This book will be a guide for professional and competitive examinations, and to undergraduate students of agriculture and agricultural engineering and master students of soil science, soil and water engineering, agricultural physics, hydrology and watershed management.

We are extremely grateful to the authors who have contributed chapters in this book. We express our thanks to Science Publishers, CRC Press for their cooperation and publication of this book.

Abrar Yousuf

Manmohanjit Singh

Contents

Chapter 1
Watershed Hydrology and Management

Anil Bhardwaj

Introduction

Water is the most abundant substance on earth upon which all life on earth is dependent. Hydrology deals with the earth's water in all its phases and is therefore a subject of great importance to society for the creation of liveable environment. Human activities such as cultivation on terraced lands, clearing of forests for different purposes, construction of roads, mining, over exploitation of groundwater, dumping wastes into rivers and reservoirs, and application of high fertilizer doses for achieving higher yields, etc., changes the pattern of distribution and circulation of earth's water. As every inhabitant living on the earth belongs to a particular watershed, they are continuously influencing quantity/availability and quality of water by their actions, and the use of water. The protection, conservation, and management of water resources and water quality depend upon all of us understanding the basic concepts of hydrology as well as that of watershed and watershed health. To do so the understanding of the hydrologic cycle is very important.

Hydrologic Cycle

We know that earth's water is always in movement. The natural water cycle that describes the continuous movement of water on, above, and below the surface

Department of Soil and Water Engineering, Punjab Agricultural University, Ludhiana, India.
Email: abhardwaj@pau.edu

of the earth is known as the *hydrologic cycle*. There is an endless circulation of water on the earth, linking oceans, land surface and atmosphere. The hydrologic cycle as shown in Fig. 1, describes the processes by which water moves around the globe. It begins with the *evaporation* of water from the ocean which then forms moist air masses. As moist air is lifted, it cools and water vapours condense to form clouds. Moisture is *transported* in the atmosphere by air currents towards land surfaces around the globe until it returns as *precipitation* over the earth surface. While falling over the earth a part of precipitation is *intercepted* by the vegetation and man-made structures. The water that eventually reaches the ground, a part of it may evaporate back into the atmosphere or it may infiltrate into the soil and percolate to become *groundwater*. Groundwater either seeps its way into the rivers, streams, and oceans, or is released back into the atmosphere through evaporation *and transpiration*. The balance amount of water that remains on the earth's surface is *runoff*, which flows and empties into lakes, rivers and streams and is carried back to the oceans, where the cycle starts again (Pidwirny 2006). Hence, the major components of the hydrologic cycle are: Precipitation— rain, snow, hail, sleet, dew, drizzle, fog, etc.; Evaporation and transpiration; Interception, depression storage, infiltration, percolation and seepage; Surface runoff, sub-surface runoff or interflow and groundwater or base flow; and water storage over and below the land surface including water stored in the soil profile.

From a global perspective, the hydrologic cycle can be considered to be comprised of four major systems as can be seen in Fig. 1., the Hydrosphere is the source of water, the Atmosphere is the deliverer of water, and the Lithosphere and the Biosphere are the users of water. The hydrologic cycle is indeed a natural machine run by solar energy and the gravitational forces with water as the material process. There is no gain or loss of water in the cycle. That means the total amount of water on the planet and its atmosphere remains same, but is continuously

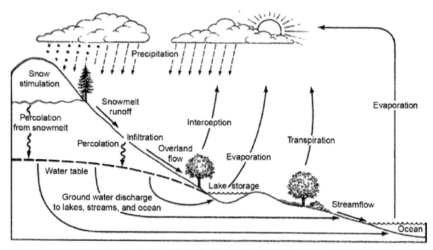

Fig. 1. Hydrologic cycle (Source: Gregory et al. 2012).

changing from one state to another and is moving at different speeds through different paths. The water falling on earth surface follows different routes on its way back to the ocean. The shortest leg of this journey is the water falling directly into the ocean. The longest leg of journey is probably the water infiltrating into the land surface and percolating down to join the groundwater, which eventually flows to the streams as spring flow and finds its way back to the ocean.

Hydrology and Watershed

The science which deals with water and its movement in the hydrologic cycle is known as *hydrology*. That means it deals with occurrence, circulation and distribution of water from the earth and earth's atmosphere. It is concerned with the water falling as precipitation on land surface, water in the streams, lakes and reservoirs, soil and rocks below the earth's surface. From an application point of view hydrology deals with occurrence, movement, distribution, circulation, storage, use, development and management of water. Traditionally, hydrology is divided into two main branches: surface water hydrology, and groundwater hydrology. Surface water hydrology deals mainly with water on the earth's surface, whereas groundwater hydrology deals with the water below the earth's surface. However, groundwater may appear as surface water or vice versa. Hydrology, which treats all phases of earth's water, is a subject of great importance for the inhabitants of the earth and their environment. Hydrologic knowledge helps us to solve water-related problems: the problems of quantity, quality and availability within a hydrologic unit known as a watershed.

A *watershed* is defined as the area above a certain point which drains water through that particular point, i.e., outlet. In other words, it is an area of land which drains or sheds all of the incoming excess precipitation at the same place, toward the same body of water or the same low elevation area resulting from its topography. This means that a watershed's boundary is defined by its topographic high points. The precipitation that falls within the boundary of a watershed would flow as excess precipitation towards its outlet, i.e., point of lowest elevation. Depending on the location of the outlet, the watershed area would be different. A watershed is fairly simple to identify in hilly areas because its boundaries are well defined by ridges. However, in flatlands such as the alluvial plains, identifying topographic high points may be quite challenging because the highest and lowest elevations may have a difference of only a few centimeters. At all points on the earth's surface, even where there is no evidence of surface runoff flow, a watershed does exist. This is because a difference in elevations exists everywhere, and when rainfall occurs, even if it is infrequent, the topographical features of the watershed will determine where runoff water will accumulate and flow. Also, all the drainage lines are located on low points on the land where surface runoff accumulates and flow. *The size of a watershed* depends on the location of the outlet and it is largest when the streams or river of that watershed discharges directly into an ocean. In that case the watershed might be referred to as a river basin and the rivers involved in such cases generally used to be perennial rivers. A river basin or a large watershed or catchment includes a number of small watersheds within its boundary, each draining runoff into the same river. Watershed or river basin boundaries do not respect district or state boundaries determined by political considerations.

Scales in Hydrology

Depending on a given hydrologic problem and the situation, the hydrologic cycle or its component processes can be assumed to vary at different scales of space and time.

Spatial scales

From the point of view of hydrologic studies, the three spatial scales are readily distinct. These are the global scale, the river basin scale and the watershed scale. The global scale is the largest scale and the watershed is the smallest spatial scale.

Hydrologic study at the *global scale* is necessary to understand the global fluxes and global circulation patterns. The global hydrologic study can be considered to be comprised of three major systems namely the oceans, the atmosphere, and the land surfaces. The principal processes that transmit water from one system to another are required to be considered. These are precipitation, runoff, groundwater and evaporation. The results of these hydrologic studies are important in water resource planning and assessment at national or regional level, weather forecasting, and climate change studies.

In the *river basin scale*, the spatial coverage can range from a few square km to thousands of square kms. In the water movements of the earth system, three systems can be recognized and considered. These are the surface system, the subsurface system, and the aquifer system. When the focus is on the hydrologic cycle of the land surface system, the dominant processes to be considered are precipitation, evaporation and transpiration, infiltration, and surface runoff. The surface system comprises of three subsystems: vegetation, topography and soil. The exchange of water among these subsystems takes place through the processes of infiltration, base flow or exfiltration, percolation, and capillary rise. These subsystems abstract and store water from precipitation through interception, depression and detention storage, which is either lost to the atmospheric system or enters subsurface system.

The watershed scale or *micro scale* is the smallest scale for conducting a hydrologic study for studying the different components of hydrologic cycle. It is more or less similar to river basin scale except the spatial coverage on the earth system. The spatial coverage of the watershed scale can range from less than a hectare to a few thousand hectares. As in the case of basin scale, three systems can be recognized to study the water movement of the watershed: the surface system, the subsurface system, and the aquifer system. The surface system of a watershed comprises of three subsystems: vegetation, topography and soil. These three subsystem characteristics are generally manipulated within a watershed to modify the response/output of the watershed in the form of runoff to different values of rainfall input. This makes watershed the most important and basic spatial scale to modify hydrologic response as per the needs of the inhabitants and the environment.

Temporal scales

The time scale used in hydrologic studies could be anything from a storm lasting for a few hours to a study spanning many years. It depends on the nature of the hydrologic problem and its objectives. Hourly, daily, monthly, seasonal or annual time scales are common. Sometimes the time interval for the collection of data determines the time scale. The time interval of the available data also affects the time scale of the hydrologic study.

A *hydrologic variable* like rainfall varies in both the time and space within a watershed. However depending on the objective or purpose of the study, type of hydrologic analysis and above all the spatial scale, rainfall can be assumed to be either constant in both time and space; constant in space but varying in time; or varying in both time and space. The spatial scale or size of the watershed determines which one of these assumptions is reasonable from a practical point of view. For *small watersheds*, rainfall can be assumed constant in both time and space in modeling rainfall—runoff relationships. As the size of the watershed increases to *medium size*, rainfall is considered variable in time but constant in space. However rainfall over a *large watershed* or a river basin is assumed to vary both in time and space. The areal extent of these watersheds may be different in hilly areas and in plain areas. Rainfall may be considered as constant over larger areas when watershed land topography is plain, as compared to hilly watersheds.

Hydrology of Small vs. Large Watersheds

The hydrology of small watersheds is different from that of the large watersheds. Small watersheds, may also be called micro-watersheds, depending upon their size, and are nested within larger watersheds. In small watersheds the headwater or upland area used to be small individually, but in large watersheds, the headwater areas used to be more. Also, well defined channels used to be less in number in small watersheds; therefore overland flow is more predominant. In large watersheds, channel systems used to be well defined, and extensive in its areal extent, hence their channel flow domination. The land use/land cover and rainfall intensity significantly affects runoff generation and runoff flow in small watersheds as compared to large watersheds.

Watershed as Hydrologic System

The watershed is a fundamental concept in hydrology and is the basis for understanding the hydrologic processes and for the planning and management of water resources. Storage and movement of water at a watershed scale is complicated due to the coupled processes which act over multiple spatial and temporal scales (Yu and Duffy 2018). Hydrologic processes within a watershed are extremely complex and are difficult to understand completely. However in an absence of perfect knowledge, these processes may be studied by means of the *systems concept* (Chow et al. 1998). Considering the watershed as a *hydrologic system* that can be defined as a structure or volume in space, surrounded by a boundary, that accepts water and other inputs, operates on them internally, and produces them as output. Here precipitation is the input, distributed in space over the watershed area; stream flow in the form of runoff concentrated at the outlet of the watershed is the output. Evaporation, transpiration and sub-surface flow could also be considered as outputs, but these are small compared to runoff or the stream flow during a storm. The structure of the system consists of the drainage lines over the watershed land surface or flow paths through the soil below the land surface and includes the tributary streams which merge to form stream flow at the watershed outlet. The schematic representation of systems operation is shown in Fig. 2. That means the watershed receives precipitation as input, operates upon it through the processes of interception, infiltration, percolation, overland flow, and channel flow, etc., and depending upon the soil, slope, and land-use/land-cover, antecedent moisture and other watershed characteristics generate output in the form of runoff as stream flow going out of the watershed through its outlet.

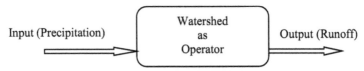

Fig. 2. Watershed as a hydrologic system.

Global vs. Watershed Hydrologic System

Water is being continuously moved between the atmosphere, the earth land surface and the oceans. This is known as the global hydrological cycle or *global hydrologic system*. This system is a closed system as there is no input or output, i.e., no water comes into or leaves the planet. On the other hand the *watershed hydrologic system* is an open system as it has a range of inputs and outputs as well as storages, transfers and flows. Inputs include precipitation including rain, snow, sleet and hail, groundwater flow from adjoining watersheds, and solar energy for evaporation. Outputs move water out of the watershed and include evaporation and transpiration from vegetation, runoff into the sea and percolation of water to underlying rock strata and into underground storages. Storages include rivers, lakes, glaciers, soil storage and groundwater storage along with water stored on vegetation, structures, etc., as interception following precipitation. Transfers or flows include infiltration, percolation, overland flow, surface flow over the land surface in streams, rivers and other drainage channels.

Watershed Hydrologic Processes

An understanding of the various hydrologic processes within a watershed is essential to keep the watershed in good health through the effective management of rainwater/runoff, soil water and soil erosion. When we consider and study the occurrence, movement, circulation and distribution of water, i.e., hydrologic cycle and its component processes within the boundaries of a watershed, it is called *watershed hydrology*. The term describes how water moves and is stored within a watershed, what are the various water inputs into the watershed and water outputs from the watershed. Understanding how water is used and cycled through a watershed provides the foundation for understanding and describing how land and water interact within that watershed. The major component processes to be studied under watershed hydrology are precipitation, evapotranspiration, infiltration, and runoff and stream flow.

Precipitation

Precipitation provides the primary input of water into the watershed. Precipitation is the moisture or water that falls from the atmosphere in the form of rain, snow, sleet, fog or hail. It varies in its amount, intensity, and form by season and the geographic location of the watershed. However rain and snowfall contribute water significantly to the watershed hydrologic system. In most parts of the world, snow and rainfall are observed and records are maintained. The watersheds located in Himalayas that are at the mean sea level of around 2000 m generally receive precipitation more as snowfall and less as rainfall, and those located in foot-hills or lower hills and plain areas including coastal regions receive that as rainfall. Precipitation is influenced by the elevation of the watershed. However, rainfall being the predominant form of precipitation causing flood flow, the term precipitation is synonymous with rainfall. Himalayan as well as the coastal watersheds tends to have higher amounts of precipitation than the plain lowlands. It may be due to the orographic effect, in which rising air mass currents cool, condense and release moisture as precipitation. The leeward side of the mountains or barriers receive less precipitation than the windward sides because most of the available moisture in the air mass is lost to precipitation before it reaches the leeward side. The snow packed watersheds contribute stream flow significantly as these act as source of water round the year for the major perennial river systems of the world. In fact the factors such as rainstorm characteristics that is; amount, duration, intensity and average return period determines whether the rain water will flow into streams or infiltrate into the ground. And this information is crucial for crop planning and management as well as for engineering design of water harvesting and flood control structures in the watershed.

The capacity of vegetated surfaces to intercept and store the precipitation water is of great practical importance to the hydrologists. A part of the precipitation while falling from the atmosphere is trapped or intercepted by the vegetation and other structures above the ground surface and is evaporated back to the atmosphere. This portion of the precipitation is known as *interception losses*, and is not available either for infiltration or for runoff generation. As such, interception and its subsequent evaporation constitute a net loss to the watershed hydrologic system which may assume considerable values under certain conditions. It may be responsible for losses reaching 10%–20% of the total precipitation, annually. Interception is a function of precipitation characteristics and the type, age and density of vegetation. The amount of interception, although negligible during the extreme events, is an important term of water balance. Interception water losses from tall forests exceed those associated with lower vegetation, such as grass land and agricultural crops. Coniferous trees tend to intercept more water than deciduous trees on an annual basis as the deciduous trees drop their leaves for a period of time (Chang 2006). The presence or absence of vegetation not only affects the amount of precipitation reaching the watershed surface, but also its kinetic energy, and thus its capacity to detach and transport soil material.

Depression storage

Depression storage is the amount of rainwater stored in the micro or macro depressions on the watershed surface before starting of runoff. The roughness of the soil surface, including roughness brought about by tillage, affects runoff and erosion, and determines the volume of water that can be held on the surface as depression storage. Four grades of surface roughness are categorized (0–1.2 cm; 1.2–2.0 cm; 2.0–3.0 cm; and > 3 cm micro-relief) in relation to tillage practices (Morgan et al. 1998). Only a proportion of the depression depth constitutes effective depression storage. It produces a rainwater loss that depends on the local characteristics of slope, land use and soil type. The amount of water that is stored in the surface depressions is ponded to evaporate or be infiltrated later. Hence, the rainwater loss then depends on the evaporation from the water surface and the infiltration. It has been found that no appreciable surface storage exists on slopes above 20%.

Evapotranspiration

Evapotranspiration is a loss of water from the watershed hydrologic system unlike precipitation, which is an input to the system. Evapotranspiration is the combined net effect of the processes of evaporation and transpiration. *Evaporation* is a loss of water from the land surface and water bodies, and *transpiration* is the evaporation of water from leaf stomata following movement of soil water and ground water from the roots upward through the plants and trees. Transpiration accounts for approximately 10% of all evaporating water. The evaporation and transpiration depend on the same physical processes to transform water from a liquid to a gas and both the processes result in a loss of water from the watershed. Hence these processes are often considered together and termed as evapotranspiration. This process moves large quantities of water from the soil and land surface back to the atmosphere. More than 50% of the total amount of precipitation is returned to the atmosphere through this process. Evaporation and transpiration rates vary widely depending upon many factors, including precipitation, temperature, aspect, humidity, and wind speed (Gregersen et al. 2007). Higher temperatures usually result in increased evaporation and transpiration unless soil moisture is limited. Aspect, i.e., the position of watershed slopes or land surface relative to the sun, affects the amounts of solar radiation and heat received with the result that both evaporation and transpiration increase from north to east to west to south-facing aspects. Lower relative humidity also increases evapotranspiration because dry air has a greater capacity to accept moisture than more humid air at same temperature. That is why evaporation and transpiration during precipitation events used to be very low as the air is saturated with moisture. Evaporation increases in response to wind because it energizes the change from liquid water to water vapour at the molecular level, and also because moist air is moved away from the water source and replaced with relatively dry air. Similarly, when plants transpire, a thin layer of air around the leaves becomes saturated. Wind stirs and moves that saturated air away from the leaves and replaces it with drier air that enhances evaporation from stomata. The amount of evapotranspiration from an area under field crops, in addition to these factors depends on how much water is available in the root zone, which depends on the field capacity of soil. In forests, transpiration accounts for much greater loss of water than any other mechanism or process in the watershed. A mature tree can transpire tens to hundreds of litres of water per day, depending upon soil moisture availability.

Infiltration

Infiltration is the entry of precipitation water into the soil surface within the watershed. It ensures that moisture will be available to sustain the growth of vegetation and helps to sustain the ground water supply to wells, springs and streams. The rate of infiltration may be influenced by the watershed characteristics such as soil type, antecedent moisture content of soil, slope, land use and land cover, and also the precipitation characteristics like amount, intensity and duration. On reaching the ground surface, water infiltrates into the soil, saturates the soil in the crop root zone and percolates down to the groundwater reservoir or it may flow over the land surface as runoff. Percolation is the infiltration process below the root zone. Light textured soils having large well-connected pores tend to have higher infiltration rates than heavy textured soils. Land use and land cover also affects infiltration. Infiltration would be higher for soils under forest vegetation and trees than bare land. Tree roots loosen the soil and provide flow paths for infiltrating water. Crop foliage and residues and also surface litter reduce the impact of falling rain drops and prevent choking of the soil pores and passages, thereby maintaining infiltration rates.

Surface runoff or overland flow

Surface runoff is the rainwater that travels over the land surface in the watershed towards the stream channel after satisfying all the precipitation losses. It is generated either when rainfall intensity exceeds the infiltration capacity of the soils or where the soil is already saturated from below. Runoff generated in first case is called infiltration excess runoff or Hortonian overland flow, and that in second case it is called saturation excess runoff or saturation overland flow. Hortonian overland flow is an important runoff mechanism in arid and semiarid regions, where rainfalls tend to be intensive and infiltration rates low or in urban areas having impervious surfaces. Saturation excess runoff mechanism of surface runoff generation occurs primarily on the lower slopes of the watershed and along valleys adjacent to stream channels. Subsurface runoff, also called interflow represents that portion of infiltrated rainfall that moves laterally through the upper soil layers until it reaches the stream channel. Interflow moves below the ground surface but above the water table. The movement of interflow or subsurface runoff is much slower than surface runoff. The proportion of total runoff that moves as subsurface runoff depends on the spatial and temporal characteristics of rainfall and physical characteristics of the watershed. Generally, mechanized agricultural lands and forest lands or land having thin soil layer overlying more impermeable soil layers tend to promote more interflow. In sloping situations, particularly if there is a reduction in permeability with depth, lateral flow develops in both the litter layer and the upper soil zone. Water then flows laterally down the slope as interflow, meeting and becoming part of stream flow. Base flow is different from interflow in the sense that it responds much more slowly to rainfall and does not fluctuate rapidly. It represents the drainage of water to the stream from deep groundwater.

Infiltration water percolates below the root zone and recharges the groundwater that provides water for stream flow through contributions known as *base flow*. Base flow is the portion of stream flow that is not attributed to the current precipitation; it may be rainfall or snowmelt inputs and is the only portion of stream flow that is present during precipitation-free periods. Base flow occurs at the existing intersection between the water table and the bed of stream channels. On the other hand *storm flow* is the component of *stream flow* that results directly from current precipitation events. Surface runoff and interflow constitute storm flow in a watershed. In forest watersheds, in the beginning of a precipitation event only small portions of the watershed area actively contribute to storm flow. But contributing areas continuously expand non-uniformly throughout the event. Most of the areas that contribute directly to storm flow tend to be adjoining to the stream channels where soils already have higher antecedent soil moisture contents, and areas with shallow soils that become saturated rapidly and therefore can release water for stream flow quickly. In agricultural watersheds, stream flow generation occurs quite differently. Due to tillage operations in agricultural systems a till layer that is more compacted than the overlying soil often forms just below the depth of the tilling implement (15–20 cm) below the land surface. This till layer retards the downward movement of infiltrating water and diverts it laterally at this shallow depth. Consequently, precipitation water moves laterally as subsurface flow at a relatively rapid rate and becomes a part of stream flow quickly. In urban watersheds, there is much less opportunity for precipitation to infiltrate into the soil because of large impervious concrete surfaces. As urban runoff flows through drains directly to streams and rivers, stream flow increases spontaneously (Pamela et al. 2015).

Watershed Hydrologic Budget

The hydrologic cycle can be understood quantitatively, in the form of a hydrologic budget. Watershed hydrologic budget is in fact a mathematical statement of the components of the hydrologic cycle within a watershed, estimated by considering the input or inflow of water to the watershed, output or outflow of water from the watershed, and the total change in the amount of water in storage within the watershed, over a specified period of time. It is represented by the following equation:

$$P-R-ET = \Delta(S_s + S_{GW}) \tag{1}$$

Where, P is precipitation, R is stream flow passing through the watershed outlet, ET is evapotranspiration, $\Delta(S_s + S_{GW})$ is the change in soil moisture plus groundwater storage within the watershed. Both volume as well as depth units can be used to measure the variables in the above equation. Negative terms (R & ET) on the left hand side of the Eq. (1) indicate outflow of water from the watershed and positive terms indicate inflow of water into the watershed. The term on the right hand side can either be positive or negative, depending upon increase or decrease in the net total moisture storage in the soil profile and groundwater reservoir of the watershed during the periods of interest. Stream flow will occur only if inputs of water to the watershed exceed all of the other outputs or uses of water in the watershed. If other water demands or outputs exceed total inputs, stream flow will not be there. Runoff water which is going as stream flow is required to be managed to a minimum within the watershed through conservation treatments to maintain watershed hydrologic health.

For application of hydrologic budget equation on a watershed, the variables of the equation are required to be determined. Precipitation is measured by rain or snow gauges; runoff by various devices such as weirs, flumes, current meters and depth gauges; infiltration by infiltrometers, or estimated through rainfall simulators or precipitation–runoff data; evapotranspiration is estimated by using evaporation pans, energy budgets, heat and mass transfer methods or empirical relationship based on climatic factors. Soil moisture can be measured by using neutron probes, sensors or gravimetric methods. Groundwater storage and flow is exceedingly difficult to estimate since knowledge of the geology of the watershed is essential. The evaluation of the storage terms depends on the time period over which the water balance is computed. On an annual basis, the change in water content of the root zone is likely to be small in relation to the total water balance and can be neglected. Over a shorter period, the change in the soil water storage can be significant and must be considered. When applied carefully, the hydrologic budget equation can yield good estimates of the magnitudes of required hydrologic variable.

Considering that hydrology is not an exact science, reasonable well-founded assumptions are required to solve practical problems in the field. The hydrologic budget equation when applied to *large watersheds*, the assumption that groundwater inflow and outflow across the boundary of the watershed is zero is quite valid as generally groundwater divide follows surface divide. Also when considering a large time span such as an year, the term $\Delta(S_S + S_{GW})$, i.e., the change in soil moisture plus groundwater storage within the watershed becomes approximately zero. Hence the hydrologic budget equation becomes:

$$P - R - ET = 0 \tag{2}$$

Considering the hydrologic budget equation for *individual storms*, the amount of ET is much smaller and the change in sub-surface storage is due to infiltration (I_N) such that $\Delta S_{GW} = I_N$; and the equation reduces to:

$$P - R - I_N = \Delta S_s \tag{3}$$

where, ΔS_S consists of interception and depression storage, and when it is coupled with infiltration it accounts for abstraction (A_b), i.e., $A_b = I_N + \Delta S_S$. And hydrologic equation becomes:

$$P - R = A_b$$

and

$$\tag{4}$$

$$R = P - A_b$$

This means that for an individual storm, the precipitation in excess of abstraction in the watershed would go as runoff or stream flow.

The hydrologic budget equation provides a relatively simple way for estimating the change in water availability in response to the change in different watershed management interventions. This equation in fact is used to describe and estimate or predict the hydrologic response of an entire watershed to an input precipitation as watershed is assumed to behave like a closed system such that outputs from the watershed are fully dependent only on the inputs within the watershed. The boundaries of the watershed ensure that this hypothesis holds good. However, under the situations where the transfer of water across the watershed boundaries do occur, and the magnitude of those losses or gains have not taken in to account, estimates using the hydrologic budget equation that would be erroneous.

Watershed Hydrology and Health

Watershed hydrology is driven by climatic processes; watershed characteristics such as topography, vegetation and geology; and human activities related to water and land use. Aquatic ecosystems are dependent on surface and ground water availability and supply. Rainwater that infiltrates into the soil recharges the aquifers and then becomes the source of water to springs, wetlands, streams, and lakes. These hydrologic regimes create habitat and are important to aquatic life. But the amount of water supply will vary depending on precipitation, evaporation and ground and surface water hydrology. Natural disturbance processes are critical to establishing healthy hydrologic regimes, i.e., flow in rivers, water supply to lakes and groundwater systems. Health of a watershed is indicated by its landscape conditions—configuration of natural land and vegetative cover; habitat conditions—flowing (streams and rivers) and standing (lakes, ponds, and wetlands) waters; hydrology—surface and groundwater; geomorphology—watershed characteristics; water quality—chemical and physical constituents; biological conditions—the presence, numbers and condition of aquatic organisms and communities in an aquatic ecosystem; and attributes of vulnerability—natural processes and anthropogenic influences (population and climate). All these indicators are related to the supply and availability of water in the watershed. If water supply is sufficient, both quantitatively and qualitatively, these indicators would indicate the positive health of the watershed. This shows that health of watershed is mainly dependent on its hydrology. Watershed health, including the natural environment, hydrology, water quality, and aquatic ecology, was assessed for the Han River basin (34,148 km²) in South Korea using the Soil and Water Assessment Tool (SWAT) (Ahn and Kim 2017).

Managing Watershed Hydrology

In watershed management projects, a hydrologist may be more interested in the amount of runoff or excess rainfall generated and its management within the watershed boundaries in order to keep the watershed in good health. A large watershed can be subdivided into a more detailed hierarchy of micro-watersheds of size ranging from a few square meters to a number of hectares, for enabling rainwater to infiltrate into the soil and controlling excess runoff because of high intensity storms. The amount of runoff generated due to a rain storm depends on the rainfall infiltrating into the soil, and that depends on the porosity of the soil within the watershed or its micro-watersheds. Depending upon the quantity of excess rainfall and runoff flow rates, watershed treatments on forest land, agricultural land, including that of drainage lines/streams are planned and executed for its proper management in meeting watershed water demands, and safe disposal without causing any adverse effect along its course of flow.

Some of the commonly adopted watershed treatments are trenching, bunding, terracing, water harvesting and silt detention structures, drop structures/check dams, spurs, retaining walls, etc. Construction of these interventions has been an integral and important component of most of the watershed development programmes in managing watershed hydrology. These measures perform one or more functions namely; water conservation, moderation of floods, soil erosion control, sediment control and drainage. They facilitate the establishment of vegetation; provide protection against the damaging runoff at points that cannot be adequately protected in any other way. A lot of work has been done and reported on watershed treatments and their effectiveness, around the globe, though information is scattered in the relevant literature. Some of the experiences in hilly areas are described in the following sub-sections.

The treatments such as *bunding and terracing* are relevant especially in the arable sloping lands of the watershed. While bunding is the construction of small embankments or bunds across the slope of the land, terracing is a method of modifying the land surface for soil and water conservation. Terraces may be broad base terraces and bench terraces. Bunding and terracing decrease the length of slope and thus reduce the concentration of runoff and hence control soil erosion. Bunding is suitable for lands having slopes from 2% to 10%. Hydrology of areas with less than 2% slope is controlled by using biological measures. Bench terracing is suitable for lands having a slope beyond 10%. According to a study, the average peak rate of runoff which was 2.2 times that of a contiguous reference forest watershed before the treatment of an agricultural watershed, reduced to 0.11 times after bunding and terracing. Also 2.12 times more rain water/runoff which was earlier leaving the watershed through its outlet before the treatment, came down to 0.43 times that from the referenced watershed after the treatment (Samra 2000). Wei et al. (2016) reported that terracing reduces the soil erosion and runoff by 11.5 and 2.60 times respectively and increased the biomass and soil water recharge 1.94 and 1.20 times. Kosmowski (2018) investigated the effect of two widely adopted soil water management practices, terraces and contour bunds, on yields and assesses their potential to mitigate the effects of climate change in Ethiopia. Although the yield on terraced plots was slightly lower than the non-terraced plots but it was observed that terraced plots acted as a buffer against the 2015 Ethiopian drought, while contour bunds did not. It was concluded that terraces have the potential to help the farmer deal with current climate risks.

Similarly, Gebreegziabher et al. (2009) provided evidence of a positive effect of contour bunds on water utilization and soil conservation. According to a study conducted by Adimassu et al. (2012), soil bunds brought about a significant reduction in runoff and soil loss. Plots with soil bunds reduced the average annual runoff by 28 per cent and the average annual soil loss by 47 per cent. Consequently, soil bunds reduced losses of soil nutrients and organic carbon.

Water harvesting structures (WHS) may be small earthen dams (height < 15 m) or concrete/stone dams, constructed across a seasonal stream on a narrow gorge or farm ponds, constructed at the farm level to collect and store rainwater. These WHSs not only store runoff water for providing supplemental or life-saving irrigation for managing frequent droughts in the rain-fed watersheds but also recharge ground water, control floods and siltation in the downstream reservoirs/ areas, conserve soil, act as source of water for wild animals and vegetation around the water harvesting structure, help in reclamation of gullied land downstream of the structure, act as a pond for fish culture, meet the water needs of the nearby inhabitants and their cattle etc. The site of the WHS is selected at such a place that ensured enough of an catchment area on its upstream side to provide the required volume of runoff, enough storage capacity for the runoff, provision for an emergence spillway to pass the excess runoff safely, and the command area to irrigate, as near to the site as possible (Sur et al. 1999). The cost of these structures varies depending upon their location. Studies conducted in rain fed hilly watersheds, indicates increase in cropping intensity from 170% to 200%; *rabi* crop yields doubled, increase in number of milch cattle by 45%, and increase in daily milk production by 103%, in addition to the reclamation of downstream area, income from fish culture and rise in water table (0.2 to 2 m) due to groundwater recharge, after the construction of WHS (Sur et al. 2001, Samra 2000).

Vegetation as *gully control measure* provides the cheapest way to control runoff and soil erosion, when the gully gradient is low. In situations where it becomes difficult to establish vegetation on the gully bed, because of a significant runoff flow, temporary structures are constructed to function until the vegetation becomes well established. Temporary structures are made up of locally available materials such as brush wood, woven wire, planks or loose rocks, etc., and are cheap and easy to construct. If the runoff is large and ultimate control by vegetation is not feasible, permanent structures have to be constructed. These are built of masonry, reinforced concrete or earth utilizing concrete or steel pipe spillways. However, these structures are costly and require careful engineering studies and adequate planning and design before their construction. In a study conducted on hilly rangeland watersheds, temporary structures along with vegetation and fencing of the micro-watershed reduced runoff by 31%, sediment concentration by 55% and sediment yield by 71% (Bhardwaj 2002). The immense pressure of population on the forests in recent years has resulted in deforestation causing the denudation of watershed land in many regions of the world. Enclosure against biotic interference (cattle grazing, tree felling, etc.), restores vegetative cover in these watersheds and controls generation of excess runoff. Studies indicate that even the simple enclosure of a hilly watershed to biotic interference reduces runoff from 30% to 7% of the annual rainfall (Bhardwaj 2002).

Rivers and streams while flowing have tendency to erode away their banks and change their flow paths. Protection of stream banks and slopes involves construction of revetments, training walls, spurs or retards, etc. Spurs may be of repelling, attracting or deflecting type, and constructed depending upon grade, width, depth, velocity of flow and characteristics of the material carried by flowing water. Sometimes single row or double row live spurs are being used to confine the stream flow to a slightly narrow bed. Gorrie (1946) suggested a herringbone plantation of sand loving plants such as *"nara"* (*arundo donax*) and *"banha"* (*vitex negundo*), set at a slight angle (10°–15°) to the stream flow direction. Sand gets deposited in between the rows thereby raising a new platform. Gabion spurs are ideally suited for stream flow training as they are flexible, porous, and economical. These measures are economically viable with a benefit–cost ratio 2.7 (Bhardwaj and Rana 2008).

Planning of watershed treatments should be done on the basis of the watershed and their execution should start from the head end of the watershed, proceeding the way the water flows downstream. However, the construction of these conservation structures is a costly affair, involving a huge sum of money and efforts. It requires good judgement in determining the need for them and the extent of their use. If these measures are not properly selected, designed or located on the most appropriate site, they may do more harm than good. Hence their planning and design should be based on the detailed study and surveys of the watershed.

Conclusions

Hydrologic processes govern water movement in a watershed through terrestrial environments and as groundwater and surface water. An understanding of the various hydrologic processes within a watershed is essential to keep the watershed in good health. The hydrologic budget equation provides a relatively simple way for estimating the change in water availability in response to the prevailing climate, topography, soil, and land use and covers conditions in the watershed. Health of the watershed is mainly dependent on its hydrology. Watershed health indicators are related to the supply and availability of water in the watershed. If water supply is sufficient, both quantitatively and qualitatively, these indicators would indicate the positive health of the watershed. Hence to keep a watershed healthy, watershed hydrology needs to be managed properly. Watershed treatments are required to be identified, planned and designed carefully based on the detailed study and surveys of the watershed for proper conservation and management of precipitation water to meet watershed water demands, and safe disposal. Local inhabitants must be involved in the whole process, for the sustainability of the adopted measures to keep the watershed hydrologic system, healthy.

References

Adimassu, Z., K. Mekonnen, C. Yirga and A. Kessler. 2012. Effect of soil bunds on runoff, soil and nutrient losses and crop yield in the Central Highlands of Ethiopia. Land Degrad. Dev. 25: 554–564. doi.org/10.1002/ldr.2182.

Ahn, So-Ra and S. Kim. 2017. Assessment of integrated watershed health based on natural environment, hydrology, water quality, and aquatic ecology. Hydrol. Earth Sys. Sci. Discuss. 21: 5583–5602. DOI:10.5194/hess-2017-88.

Bhardwaj, A. 2002. Controlling reservoir sedimentation through watershed treatment—a review. Indian J. Power River Valley Dev. 2: 97–100.

Bhardwaj, A. and D.S. Rana. 2008. Torrent control measures in *Kandi* area of Punjab—A case study. J. Water Manage. 16: 55–63.

Chang, M. 2006. Forest hydrology—An introduction to water and forests. Boca Raton, CRC Press.

Chow, V.T., D.R. Maidment and L.W. Mays. 1988. Applied hydrology. Int. Edition, McGraw-Hill Book Co., New York.

Gebreegziabher, T., J. Nyssen, B. Govaerts, F. Getnet, M. Behailu, M. Haile and J. Deckers. 2009. Contour furrows for *in situ* soil and water conservation, Tigray, Northern Ethiopia. Soil Tillage Res. 103: 257–264. DOI: 10.1016/j.still.2008.05.021.

Gorrie, R.M. 1946. Soil Conservation in Punjab. Govt. Printing Press, Lahore, Pakistan.

Gregersen, H.M., P.F. Ffolliott and K.N. Brooks. 2007. Integrated Watershed Management: Connecting People to Their Land. CAB International, Cambridge, Massachusetts.

Gregory, K.J., I.G. Simmons, A.J. Brazel, J.W. Day, E.A. Keller, A.G. Sylvester and A. Yanez-Arancibia. 2012. Environmental Sciences: A Student's Companion. SAGE Publications, California.

Kosmowski, F. 2018. Soil water management practices (terraces) helped to mitigate the 2015 drought in Ethiopia. Agril. Water Manage. 204: 11–16. doi.org/10.1016/j.agwat.2018.02.025.

Morgan, R.P.C., J.N. Quinton, R.E. Smith, G. Govers, J.W.A. Poesen, K. Auerswald, G. Chisci, D. Torri, M.E. Styczen and A.J.V. Folly. 1998. The European Soil Erosion Model (EUROSEM): documentation and user guide. pp. 1–89. Silsoe College, Cranfield University.

Pamela, J., W.J.W. Karl and E.S. Jon. 2015. Fundamentals of watershed hydrology. J. Contemp. Water Res. Edu. 154: 3–20.

Pidwirny, M. 2006. The Hydrologic Cycle. Fundamentals of Physical Geography, 2nd Edition. http://www.physicalgeography.net/fundamentals/8b.html.

Samra, J.S. 2000. Soil conservation and watershed management in Asia and the Pacific-India. Asian Productivity Organization, Tokyo, 123–156.

Sur, H.S., A. Bhardwaj and P.K. Jindal. 2001. Performance evaluation and impact assessment of small water harvesting structures in the Shivalik foot-hills of Northern India. Am. J. Alt. Agric. 16: 124–130.

Sur, H.S., A. Bhardwaj and P.K. Jindal. 1999. Some hydrological parameters for design and operation of small earthen dams in lower *Shiwaliks* of Northern India. Agric. Water Manage. 42: 111–121.

Wei, W., D. Chen, L. Wang, S. Daryanto, L. Chen, Y. Yang, Y. Lu, G. Sun and T. Feng. 2016. Global synthesis of the classifications, distributions, benefits and issues of terracing. Earth-ScienceRev. 159: 388–403. dx.doi.org/10.1016/j.earscirev.2016.06.010.

Yu, X. and C.J. Duffy 2018. Watershed Hydrology: Scientific Advances and Environmental Assessments. Water 10: 288. doi:10.3390/w10030288.

Chapter 2

Runoff and Rainwater Harvesting

Junaid N Khan,[1,] Rohitashw Kumar[1] and Abrar Yousuf[2]*

Introduction

Excess rainfall is the rainfall remaining after satisfying all the hydrologic abstractions such as interception, infiltration and depression storage. Excess rainfall becomes runoff and eventually streamflow. In most urban areas, the population is increasing rapidly and the issue of supplying adequate water to meet societal needs and to ensure equity in access to water is one of the most urgent and significant challenges faced by decision-makers. With respect to the physical alternatives to fulfil the sustainable management of freshwater, there are two solutions: finding an alternate or additional water resources using conventional centralized approaches; or better utilizing the limited amount of water resources available in a more efficient way. To date, much attention has been given to the first option and only limited attention has been given to optimizing water management systems. Among the various alternative technologies used to augment freshwater resources, rainwater harvesting and utilization is a decentralized, environmentally sound solution, which can avoid many environmental problems often caused in conventional large-scale projects using centralized approaches.

[1] College of Agricultural Engineering, SKUAST-Kashmir, Srinagar 190025, India.
 Emails: rohituhf@rediffmail.com
[2] Regional Research Station (Punjab Agricultural University), Ballowal Saunkhri, SBS Nagar, 144521, India.
 Email: er.aywani@gmail.com
* Corresponding author: junaidk1974@gmail.com

Rainwater harvesting, in its broadest sense, is a technology used for collecting and storing rainwater for human use from rooftops, land surfaces or rock catchments using simple techniques such as jars and pots as well as engineered techniques. Rainwater harvesting has been practiced for more than 4000 years, owing to the temporal and spatial variability of rainfall. It is an important water source in many areas with a significant rainfall but lacking in any kind of conventional, centralized supply system. It is also a good option in areas where good quality fresh surface water or groundwater is lacking. The application of appropriate rainwater harvesting technology is important for the utilization of rainwater as a water resource.

Runoff

Runoff is the portion of rainfall which flows over the land surface and reaches the watershed outlet and discharges into any stream or channel. Runoff is formed after all the initial abstractions such as interception, infiltration, depression storage are satisfied.

When it rains over a catchment area, some of the rain is intercepted by the crop canopy, a part of is it infiltrated into the soil surface and some part is retained as depression storage. After satisfying all these abstractions, the excess rainfall begins to flow over the land surface through the small channels and joins the larger/main drainage channel to reach the catchment outlet. This flow is known as runoff (Subramanya 1993).

Types of runoff

Surface runoff

It is that portion of the runoff which enters the stream immediately after the rainfall. After satisfying all the initial abstractions, if the rainfall continues with an intensity greater than the infiltration rate of the soil, this part of rainfall flows directly over the land surface as surface runoff.

Sub-surface runoff

It is that portion of the runoff which enters into the soil and moves parallel to the land surface within the soil and reappears at the surface at some other point. Sub-surface runoff is also known as interflow or quick return flow because it takes a very small time to reappear at the surface.

Base flow

It is that portion of the rainfall which infiltrates into the soil and flows through the soil layers to reach the groundwater. The rate of flow in this type of runoff is very slow, in the order of months and years. This part of the runoff is also known as groundwater flow.

Hence, the total runoff generated by a rainfall event is the sum of surface runoff (including sub-surface runoff) and base flow.

Factors Affecting Runoff

The following are factors which affect the runoff:

Climatic factors

1. Type of Precipitation
2. Rainfall Intensity
3. Duration of Rainfall
4. Rainfall Distribution
5. Direction of Prevailing Wind

Physiographical factors

1. Size of watershed
2. Shape of watershed
3. Slope of watershed
4. Land Use
5. Soil Type
6. Soil Moisture
7. Topographic Characteristics
8. Drainage Density

Climatic factors

The effect of different climatic factors on the runoff is discussed below.

Type of precipitation

The type of precipitation is an important factor affecting the runoff. Precipitation in the form of rainfall generates runoff quickly as compared to snow.

Rainfall intensity

Among the different rainfall characteristics, rainfall intensity is a very important factor for the rainfall-runoff process. The amount and peak runoff rate resulting from the rainfall event depends on the rainfall intensity. The runoff takes place only when the rainfall intensity is greater than that of the infiltration rate of the soil. Generally, high intensity rainfall events generate high runoff and vice-versa. A number of studies have been conducted to study the effect of the different rainfall intensities on the surface runoff at the field/plot scale (Huang et al. 2013, Mohamadi and Kavian 2015). As evident from results of different studies; the influence of rainfall intensity on runoff is not straightforward. On the one hand, increased rainfall intensity leads to increased runoff, due to the fact that increased rainfall intensity may bring about the formation of the soil crust, and the development of such soil crusts would reduce the infiltration (Mu et al. 2015). On the other hand, owing to the spatial heterogeneity in the infiltration characteristics of the soil surface, infiltration would increase with increased rainfall intensity and runoff might decrease (Parsons and Stone 2006).

Duration of rainfall

The runoff generated from the rainfall event is directly related to the rainfall duration. Higher the rainfall duration, the higher will be the runoff generated.

Distribution of rainfall

Rainfall distribution over the watershed affects the runoff behaviour of the watershed. The term distribution coefficient is used to express the effect of rainfall distribution on the runoff. Distribution coefficient is defined as the ratio of rainfall at a particular point in the watershed to the average rainfall over the entire watershed. In general, higher the value of the distribution coefficient, the higher the runoff and vice versa.

Direction of prevailing wind

The direction of wind also affects the runoff. If the direction of the prevailing wind is in the direction of slope, it will result in higher peak runoff in a shorter period of time.

Physiographic factors

The effect of different physiographic factors on the runoff is discussed below.

Size of watershed

Under given rainfall characteristics, the size of watershed directly affects the runoff yield. The larger watershed will produce a higher runoff as compared to a smaller watershed. However, the larger watersheds take a longer time to drain off the entire runoff to the watershed outlet, hence have the smaller peak.

Shape of watershed

The fan shaped watershed produces higher peak runoff rate as compared to the fern shaped watershed. It is because in fan shaped watershed, all the parts of watershed contribute runoff simultaneously to the outlet in less time as compared to fern shaped watershed.

Slope of watershed

The slope of the watershed affects the overland flow and velocity of the runoff. Generally, higher the slope of watershed, the higher is the peak runoff rate in the watershed and vice-versa.

Land use

The land use of the watershed has a prominent effect on the runoff. Vegetation intercepts the rain water and increases the infiltration in the soil. Further, it retards the movement of runoff over the soil surface. In barren fields, there is no interception of rainfall and no hindrance to the flow of runoff. Naharuddin et al. 2018 studied the effect of different land use systems on runoff and soil erosion in different watersheds of Indonesia. The results showed that the highest runoff was generated from the non-agro–forestry land use system, followed by teak tree and cocoa based agro-forestry system.

Soil type

The coarse textured soils such as sandy soils have a high infiltration rate and hence produce a lesser amount of runoff. Fang et al. 2015 conducted rainfall simulation experiments to study the effect of rainfall intensity and slope gradient on runoff, soil loss and rill development under two different loess soils (Anthrosol and Cakcaric Cambisol). It was observed that runoff and soil loss from the Anthrosol soils were generally higher than those from the Calcaric Cambisol soils.

Soil moisture

The amount of soil moisture at the time of rainfall influences the runoff yield. High soil moisture means that there is less infiltration in the soil and consequently high runoff. On the other hand, when the soil is dry, infiltration is more and hence less runoff.

Topographic characteristics

The term topographic characteristics mean the undulating nature of the watershed. Usually undulating watersheds produce more runoff than the flat lands due to the slope of the watershed.

Drainage density

The drainage density is defined as the ratio of the length of all the channels in the watershed to the total area of the watershed. Its dimensions are [L^{-1}]. Higher the drainage density of the watershed, the higher is the peak runoff rate of the watershed. Ogden et al. 2011 in their study on a 14.3 km² watershed located in Maryland, USA, observed that with an increase in drainage density, particularly increases in density from low values, produces significant increases in the runoff peaks.

Estimation of Runoff

Rational method

The rational method uses existing rainfall data and land use to estimate peak runoff from small drainage areas that are less than 15 km² (Ramser 1972). The rational method used the following equation to estimate the peak runoff rate from the watershed.

$$Q = \frac{CIA}{360}$$

where,

Q is the peak runoff rate (m³/s)

C is the runoff coefficient, which is defined as the ratio of depth of runoff produced to the rainfall occurred. Its value ranges from 0 to 1

I is the intensity of rainfall (mm/h) for a duration equal to time to concentration of watershed and a given recurrence interval

A is the area of watershed (ha)

The rainfall intensity (I) is read from Intensity-Duration-Frequency (IDF) curves.

Time of concentration: Is defined as time (in minutes) taken by runoff from the remotest point of the catchment to reach to the outlet. A number of empirical equations have been developed to estimate the time of concentration. Kirpich (1940) developed the following equation to estimate the time of concentration:

$$Tc = 0.0195L^{0.77} \, S^{-0.385}$$

where

Tc is the time of concentration (minutes)
L is the length of channel (m)
S is the average slope of channel (m/m)

Haan et al. (1982) developed the following equation to calculate the time of concentration for a small watershed where the overland flow is predominant. He introduced the term for the overland flow. The equation is as follows:

$$T_c = 0.0195L^{0.77}S^{-0.385} + \left[\frac{2L_o n}{\sqrt{S_o}}\right]^{0.467}$$

where

L_o is the length of overland flow
n is the Manning's roughness coefficient
S_o is the slope of the land surface

SCS curve number method

SCS Curve Number method was developed by United States Soil Conservation Services (US SCS 1972). This method gives the depth of runoff (Q) generated by the rainfall event (P). This method is based on the potential retention capacity (S) of the watershed. According to this method, runoff is calculated using the following equation:

$$Q = \frac{(P - 0.2S)^2}{P + 0.8S}$$

where

Q is the runoff depth (mm)
P is the rainfall (mm)
S is the retention parameter (mm)

The retention parameter, S, is given as:

$$S = \frac{25400}{CN} - 254$$

where

CN is the curve number

This method introduces the curve number (CN) which describes the runoff generating capacity of the surface. The value of the curve number varies between 0–100.

The curve number method is adaptable and widely used for runoff estimation. This method takes into consideration important properties of the watershed, especially soil permeability, land use and antecedent soil water conditions. The curve number method has been used worldwide for the estimation of surface runoff from watersheds having different characteristics (Xiao et al. 2011, Bofu 2012, Ajmal 2015, Satheeshkumar 2017, Soulis 2018).

Creager's method

Creager (1945) proposed the following the empirical equation to estimate the peak runoff rate from the watersheds

$$Q_m = C_1 \times (0.386A)^{0.894 \times (0.386A)^{-0.048}}$$

where

Q_m is the Maximum or peak flow for a given return period.
A is the Catchment area (Sq. Km).
C_1 is the Creager's number (max. 130)

Inglis and De Souza formula

Inglis and De Souza (1929) established a relationship between the annual rainfall (P) and annual runoff (R) based on data from 53 runoff gauging stations. The relationship is given as follows:

For mountain ranges:

$$R = 0.85P - 30.5$$

For plateaus or plains in between 2 mountain ranges

$$R = \frac{1}{254}P(P - 17.8)$$

where

R is the annual runoff (cm)
P is the annual rainfall (cm)

Talbot Method

Runoff can be related to the morphological properties of the catchment. Runoff (Q), in m³/sec, can be calculated using the following formula

$$Q = aCA^n$$

Khosla's formula

Khosla (1960) developed an empirical relationship between monthly runoff and monthly rainfall.

$$R_m = P_m - L_m$$

and

$$L_m = 0.48T_m \text{ for Tm} > 4.5°C$$
$$L_m = 2.17 \text{ at } 4.5°C$$
$$L_m = 1.78 \text{ at } -1°C$$
$$L_m = 1.52 \text{ at } -6.5°C$$

where

R_m is monthly runoff (cm)
P_m is monthly rainfall (cm)
L_m is monthly losses (cm)
T_m is mean monthly temperature (°C)

Ryves formula

Ryves developed an empirical relationship between the peak flow rate and the watershed area in 1884. It is given as:

$$Q_p = C_r A^{2/3}$$

Q_p is the peak flow rate (m³/s)
C_d is the Ryves's constant, value ranges from 6 to 30
A is the watershed area (km²)

Dickens formula

Dicken's formula was developed in 1865 to estimate the peak flow rate from the watershed and it is is given as:

$$Q_p = C_d A^{3/4}$$

where

Q_p is the peak flow rate (m³/s)
C_d is the Dicken's constant, value ranges from 6 to 30
A is the watershed area (km²)

Direct Measurement of Runoff

Crest-stage gauges

The crest gauge is designed to measure peak discharge in a channel reach during a flood event (Murthy 2013). A crest gauge consists of an ordinary staff gauge of sufficient width, to fit into a 2-inch galvanized pipe. This galvanized pipe is fitted with threaded pipe caps on either end. The holes of approximate diameter 0.25 inch are drilled in the galvanized pipe. The pipe is the installed vertically in the channel such that the bottom of the pipe is at the datum or bed of the stream. The staff gauge is inserted at the top of the pipe along with about a capful of ground cork and the top ventilated pipe cap is replaced. When the water flows in the channel, water enters through the lower holes and rises in the pipe and carries the ground cork with it. At the highest stage, the cork adheres to the wetter staff. As the water recedes, a visual record of the highest stage is indicated by the cork adhering to the staff. The operator only needs to remove the stall, read the high stage and wipe the staff clean for additional use (http://ecoursesonline.iasri.res.in).

Staff gauge

A staff gauge is the simplest device to measure the river stage. It is usually installed vertically or may be at an angle with the vertical. The staff is rigidly attached to a permanent structure such as a bridge, pier, wall abutment, etc. The gauge indicates water-surface elevation on a staff that is graduated with clear and accurate markings in tenths of a foot or in centimetres. A portion of the scale is immersed in the water at all times (http://ecoursesonline.iasri.res.in).

Flow measurement by weirs

A weir is a barrier across a river designed to alter the flow characteristics. In most cases, weirs take the form of a barrier, smaller than most conventional dams, across a river that causes water to pool behind the structure and allows water to flow over the top. Weirs are commonly used to alter the flow regime of the river, prevent flooding, measure discharge and help render a river navigable (Arora 1980). Weirs are structures consisting of an obstruction across the open channel with a specially shaped opening or notch. The weir results in increase in the water level, or head, which is measured upstream of the structure. The flow rate over a weir is a function of the head on the weir.

The relation of flow rate over the weir to the head is used to figure out the discharge. The procedure for measurement of runoff though a weir is as follows:

1. Measure the size of weir, accurately.
2. Insert the weir into the hydraulic bench and fit it tightly.

3. Turn on the pump and open the valve, wait until water discharge over a weir. Then, close the valve and turn the pump off and allow water to drop until water flow over the weir stops.
4. Be sure that the water surface is in the same level as the weir crest or the lower tip of weir. Adjust the Hook gauge to touch the water surface. Set and record the reading to be the zero gauge reading, so that the bottom of the notch is taken as the datum.
5. Turn on the pump and open the valve again.
6. Adjust the Hook gauge to touch water surface. Read the scale as the Gauge reading and minus it by the Zero gauge reading to get the water height, H. Record H in your data sheet.
7. Measure the discharge by the Weight time measurement method. Record Q in your data sheet.
8. Adjust the valve again to get a total of 8 points of data for each type.

Weirs are mainly classified as (1) Sharp crested weirs and (2) Broad crested weirs

Sharp crested weirs

Sharp Crested weirs have the sharp-edged crest which allows runoff to flow through it. These have a sharp upstream edge to allow the water to flow freely over it without touching the front of the blade. Sharp crested weirs are of three types depending on their shapes.

Rectangular weirs: The rectangular weirs are most commonly used to measure the runoff from the small watersheds. Flow through a rectangular weir can be expressed as

$$Q = 0.0184LH^{1.5}$$

where

Q is the flow rate (lps)
H is the head on the weir (cm)
L is the width of the weir (cm)

Triangular Weirs or V notch: These weirs are generally used to measure the low volumes of the runoff. They are made with different angles like 45°, 90° and 120°. Among, the three, 90°-notch is most commonly used.

For a triangular or v-notch the flow rate can be expressed as:

$$Q = \frac{8}{15}\sqrt{2g}\,\tan\frac{\theta}{2}H^{2.5}$$

where

Q is the flow rate (m³/s)
g is the acceleration due to gravity
θ is the v-notch angle (degrees)
H is the head (cm)

For 90° V notch, the formula for discharge is given as:

$$Q = 0.0138H^{2.5}$$

Trapezoidal weir: Trapezoidal weir having the side slope of 1:4 is known as Cipolleti weir (named after Italian engineer, Cipolleti). The following equation is used to measure the discharge through the weir:

$$Q = 0.0186LH^{1.5}$$

Broad crested weir

Unlike sharp crested weirs, broad crested weirs have the broad crest. The discharge through this type of weir depends upon the shape of the weir and the crest. They are usually calibrated in the field by current meter measurements or in the laboratory by some model tests.

Flumes

In addition to weirs, the flumes are also used for runoff measurements in the field. The principle of the flumes is based on the concept of the specific energy and critical flow in open channels. Two types of flumes are commonly used: (1) Parshall flume (2) H flume.

Stage Level Recorder

A stage recorder is a device for producing a graphical, digital, or punched tape record of the temporal variation in water surface elevation. The instrument consists of a gauge height component and a time component. A float, manometer, or pressure transducer provides the gauge height component. The time element is controlled by a clock or digital timer driven by electricity, spring, or a weight.

For some hydrogeologic studies, frequent and uninterrupted water-level measurements may be needed to identify the unique properties of the groundwater flow system. In studies in which a more complete picture of water-level fluctuations is needed, automatic float-activated water-level recorders can be installed. Float-activated recorders sense changes in water level by the movement of a weight-balanced float that is lowered into the well. The stage level recorder for measurement of runoff is shown in Fig. 1.

Fig. 1. Digital stage level recorder installed for measurement of runoff at Regional Research Station, Ballowal Saunkhri.

Operational instructions

A wire attached to the float passes over a pulley on the recorder and a counterweight is attached to the other end of the wire and hangs in the well. When the clearance between the float and the well casing is small, the float cable should be set so that the counterweight does not have to pass the float, but is always above or below the water level. If the counterweight is immersed below the water level, a little extra weight should be added to offset the water's buoyancy.

Chart or graphic recorder is the simplest device, but it is not commonly in use. It is a drum chart that is actuated mechanically by a float that follows the water level. The graphic recorder provides a continuous pen and ink trace of the water level on a chart, which is graduated to record both water level and time. Battery operated clocks for graphic recorders can be set to record a wide variety of intervals, ranging from a few hours to 1 month. Data is retrieved by changing the paper chart. Now a days, digital water level recorders are available to measure the surface runoff. These recorders can record and store the data for different time intervals. The recorded data can be easily downloaded in MS Excel file on the computer using the data shuttle.

Current Meter

A current meter is oceanographic device for flow measurement by mechanical (rotor current meter), tilt (Tilt Current Meter), acoustical (ADCP) or electrical means.

When the meter is lowered in water and when it faces the current of water in the channel the wheel rotates. To keep the meter facing the direction of flow a tail is attached. This tail aligns the meter in the direction of flow. The meter is also fitted with a streamlined weight (fish weight) which keeps the meter in a vertical position. The rate of rotation of the wheel depends on the velocity of flow. A dry battery is kept on the shore or in a boat and an electric current is passed to the wheel from it. A commutator is fixed to the shaft of the revolving wheel.

It makes and breaks the contact in an electrical circuit at each revolution. An automatic revolution counter is kept in the boat or on the shore with the battery which registers the revolutions. When an electric circuit is broken, an electric bell in the boat rings or a head phone in the boat buzzes. Then the time taken for a required number of revolutions may be noted. The velocity of flow can be read from a rating table. The rating table is always provided with the meter.

Rain Water Harvesting Techniques

Following are traditional techniques are used for augmentation of water resources.

1. Small Ponds
2. Dammed Ponds
3. Cemented/Stone
4. Construction of "Bauwris"

New Techniques

1. Runoff harvesting-short-term storage
2. Semi-Circular Hoops
3. Trapezoidal bunds
4. Nala Bandhan" (Mini earthen check dams)
5. Off-contour bunds or graded bunds
6. Rock catchment
7. Ground Catchment

Rainwater Harvesting in Farm Ponds

Rainwater harvesting in farm ponds is used for water harvesting/storage structure in arable land. Farm ponds are of different types:

1. Embankment type
2. Dug out type

An embankment type pond is built across the stream in areas of gentle to moderately slope (Das 2002). Dug out type ponds are constructed by excavating the soil, in relatively level areas. For the harvesting of rainwater, farm ponds play a crucial role in the Himalayan region, and harvested rainwater is used the main source of water for irrigation, as well as for drinking, in horticulture, and agro-forestry. The solution to the water scarcity problem in the Himalayan region lies to a great extent in the rejuvenation of farm ponds. The farm pond needs scientific assessment of storage capacity and related hydro-geomorphic characteristics. In addition, earthen farm ponds as well as poly lined farm pond of different capacities are most suitable and viable for hilly areas. The construction cost of poly lined farm ponds are nearly @0.50 paise/litre storage for the drinking water and maybe used as well in the lean season for life saving irrigation in vegetable and fruit crops.

Design of Farm Ponds

The following parameters are considered for the design of a farm pond:

1. Site selection
2. Capacity of pond
3. Design of embankment
4. Design of mechanical spillway
5. Design of emergency spillway
6. Provision for seepage control

Site selection

The selection of suitable site for the farm pond is important. Following points should be considered while selecting the farm pond:

1. The site should be such that all the runoff from the catchment is concentrated towards the site.
2. The site should be such that a large capacity is obtained with the least amount of the earth work. This will make construction of pond economical.
3. The pond should be located nearest to the area where the harvested water will be used.
4. The site should be such that it provides the proper spillway for the safe disposal of the excess water.

5. The soil at the site should be impervious enough to prevent the seepage through the pond area.
6. The pond site may be surrounded by tall trees to reduce the evaporation form the pond area.

Capacity of pond

The capacity of the farm pond depends primarily on the catchment area, volume of water required and soil characteristics. The amount of water that can be harvested in the pond directly depends on the catchment area.

The ponds capacity is determined by studying the contour map of the catchment area of the pond. From the contour plan of the site, the capacity is computed for different stages using the trapezoidal or Simpson's formulae. The area enclosed by each contour is measured with the help of planimeter.

Design of embankment

The design of embankment consists of foundation, cross section and side slope. The data required for the design of embankment includes hydrologic data, climate data, geologic data and data for dams.

Foundation

The foundation should be such that it provides stable support and resistance to the seepage of water. A mixture of coarse sand and fine texture soil like gravel-sand-clay mixture, sand-clay mixtures and sand-silt mixtures are good foundation materials.

Cross section

The cross section of the embankment depends both on the nature of foundation and fill materials. The materials used for embankment construction should be fine and impervious. If the fine and impervious material is not available, then an impervious core and a cut-off trench should be provided in the embankment for seepage control.

Side slope

The side slope of the embankment depends on the height of dam, nature foundation material and nature of fill material. For sandy loam soils, the side slope of 3:1 and 2.5:1 is usually provided on upstream and downstream side respectively. Similarly for clay soil, it should be 2.5:1 on upstream side and 2:1 on downstream side.

Design of mechanical spillway

The purpose of the mechanical spillway is to dispose of the excess water in a controlled manner. The kind of spillway to be provided depends on the catchment area of the pond. Vegetative spillways are provided for ponds having a catchment area of less than 4 ha. Mechanical spillways are provided for catchment of more than 12 ha. Generally, drop spillway and drop inlet spillway are provided as a mechanical spillway in the embankment type ponds. A drop spillway handles a larger discharge than drop inlet spillway. The drop inlet spillway is constructed as a simple pipe outlet having a control valve to regulate the flow of water.

Design of emergency spillway

The emergency spillway is provided to prevent the overtopping of the pond due to unexpected inflow into the pond. It is provided at one end of the embankment such that its bottom elevation is set at a maximum expected water level in the pond. The recommended side slope of the emergency spillway is 2:1.

Seepage control

To control the seepage from the ponds, suitable lining material should be used. Most common lining material is the UV stabilized polythene sheets because of the lesser cost in material. In addition to polythene sheets, concrete lining is also applied on the pond. The concrete lining is more expensive than the polythene sheets (Fig. 2).

Design of dug out type farm pond

The following different steps are used for design of LDPE farm pond

1. Calculate the runoff volume (V1) from catchment area (A).

$$V_1 = A \times d$$

   where
   d is runoff depth, i.e., some % of rainfall.
2. Calculate design runoff volume (V), i.e., some % of total runoff volume (V1).
3. Side slope (z:1) of farm pond: The side slope of the pond depends on the type of soil, e.g., for red soil, recommended side slope is 1.5:1 and for black soil, it is 2:1.
4. Depth (d) of the farm pond can be assumed according to farm pond capacity, it should not more than 3 m.
5. The bottom width (b) of the pond can be calculated using the formula,

$$b = \frac{\sqrt{3V - d^3 Z^2}}{\sqrt{3d}} - dz$$

6. Similarly, top width (T) is calculated using the formula,

$$T = b + 2dz$$

7. Capacity of the farm pond can be determined by trapezoidal rule or Simpson's rule.

Trapezoidal formula

$$V = \frac{d}{2}\left[(A_1 + A_L) + 2(A_2 + A_3 + A_4 + \ldots)\right]$$

Simpson's formula,

$$V = \frac{d}{3}\left[(A_1 + A_L) + 2(A_3 + A_5 + A_7 + \ldots) + 4(A_2 + A_4 + A_6 + \ldots)\right]$$

where

$A_1, A_2, A_3, \ldots, A_L$ are areas of first, second, third, and so on last contours
d is vertical interval of contours.

8. Volume of excavation for the pond is calculated by prismoidal formula

$$V = \frac{(A + 4B + C)}{6} \times D$$

where

A is the area of excavation at the ground surface.
B is the area of excavation at the midway depth of pond.
C is the area of excavation at the bottom of pond.

Fig. 2. Concrete lined farm pond at Regional Research Station, Ballowal Saunkhri.

Conclusions

Quantification of runoff is very important for design of soil and water conservation structures. The quantification of runoff can be done by direct or indirect methods. Direct methods involve measurement of surface runoff from small watersheds with the help of weirs, flumes and stage level recorders. The weirs and flumes measure the discharge from the watersheds in terms of head (depth of water flowing over the weirs or flumes). The head is then converted into the discharge with the help of formulae for different weirs and flumes. Indirect methods involve application of different empirical formulae for estimation of surface runoff. SCS Curve number method, Rational formula, Creagor's method, Ryves formula are some of the methods discussed in this chapter. The erratic and uneven distribution of rainfall both spatially and temporally, necessitates rainwater harvesting to increase and sustain agricultural productivity. Different types of rainwater structures have been developed over the years. The most common rainwater harvesting constructed is the farm pond. The farm ponds should be designed carefully keeping in view the rainfall of the region and catchment area of the pond. The locally adoptable low-cost technologies for rainwater harvesting can be implemented as a viable alternative to conventional irrigation and drinking water supply schemes considering the fact that any land anywhere can be used to harvest rainwater. The Government and local communities have to identify it as an effective measure to combat the problem of finding a workable technology option for the mitigation of droughts, preserving the groundwater reserves, hindering soil erosion, and providing a dependable source of drinking as well as irrigation water. Mitigation and adaptive measures are needed to offset any future impact of climate change on agriculture and water resources.

References

Ajmal, M. and T. Kim. 2015. Quantifying excess stormwater using SCS-CN–based rainfall runoff models and different curve number determination methods. Journal of Irrigation and Drainage. 141(3). doi:10.1061/(ASCE)IR.1943–4774.0000805.

Arora, K.R. 1980. Fluid Mechanics, Hydraulics and Hydraulic Machines. Standard Publishers, New Delhi.

Bofu, Y. 2012. Validation of SCS Method for Runoff Estimation. J. Hydrol. Engg. 17(11): 1158–1163. DOI: 10.1061/(ASCE)HE.1943-5584.0000484.

Creager, W.P., J.D. Justin and J. Hinds. 1945. Engineering for Dams. Vol. 1. John Wiley, New York, USA.

Das, G. 2002. Hydrology and Soil Conservation Engineering, Printce Hall of India Private Limited, New Delhi, India.

Fang, H., L. Sun and Z. Tang. 2015. Effects of rainfall and slope on runoff, soil erosion and rill development: an experimental study using two loess soils. Hydrol. Process. 29: 2649–2658.

Haan, C.T., H.P. Johnson and D.L. Brakensick. 1982. Hydrologic modeling of small watersheds. ASAE Monograph No. 5, ASAE.

http://ecoursesonline.iasri.res.in Last access in 23 August 2019.

Huang, J., P. Wu and X. Zhao. 2013. Effects of rainfall intensity, underlying surface and slope gradient on soil infiltration under simulated rainfall experiments. Catena. 104: 93–102. http://dx.doi.org/10.1016/j.catena.2012.10.013.

Inglis, C.C. 1929. Technical Paper No. 30, PWD, Bombay.

Khosla, A.N. 1960. Silting of Reservoirs, CBIP, New Delhi.

Kirpich, Z.P. 1940. Time of concentration of small watershed. Civil Engg. 10: 362.

Mohamadi, M.A. and A. Kavian. 2015. Effects of rainfall patterns on runoff and soil erosion in field plots. Int. Soil Water Conserv. Res. 3: 273–281. http://dx.doi.org/10.1016/j.iswcr.2015.10.001.

Mu, W., F. Yu., C. Li., Y. Xie, J. Tian, J. Liu and N. Zhao. 2015. Effects of rainfall intensity and slope gradient on runoff and soil moisture content on different growing stages of spring maize. Water. 7: 2990–3008. doi:10.3390/w7062990.

Murthy, V.V.N and M.K. Jha. 2013. Land and Water Management Engineering. Kalyani Publishers, New Delhi.

Naharuddin, Rukmi, R. Wulandari and A.K. Palaloang. 2018. Surface runoff and erosion from agroforestry land use types. The J. Animal. Plant. Sci. 28: 875–882.

Ogden, F.L., N.J. Pradhan, C.W. Downer and J.A. Zahner. 2011. Relative importance of impervious area, drainage density, width function, and subsurface storm drainage on flood runoff from an urbanized catchment. Water Resour. Res. 47: W12503. https://doi.org/10.1029/2011WR010550.

Parsons, A.J. and P.M. Stone. 2006. Effects of intra-storm variations in rainfall intensity on interrill runoff and erosion. Catena. 67: 68–78. DOI 10.1016/j.catena.2006.03.002.

Ramser, C.E. 1972. Runoff from small agricultural areas. Journal of Agricultural Research 34: 797–823.

Satheeshkumar, S., S. Venkateswaran and R. Kannan. 2017. Rainfall–runoff estimation using SCS–CN and GIS approach in the Pappiredipatti watershed of the Vaniyar sub basin, South India. Model. Earth Syst. Environ. 3: 24. DOI: 10.1007/s40808-017-0301-4.

Soulis, K.X. 2018. Estimation of SCS curve number variation following forest fires. Hydrol. Sci. J. 63: 1332–1346. https://doi.org/10.1080/02626667.2018.1501482.

Subramanya, K. 1993. Engineering Hydrology, Tata Mc Graw Hill Co. Ltd. New Delhi.

US Soil Conservation Service. 1972. Hydrology: National Engineering Handbook, Section 4, Washington DC.

Xiao, B., W. Hai., F. Jun., H. Peng and D. Hou. 2011. Application of the SCS-CN model to runoff estimation in a small watershed with high spatial heterogeneity. Pedosphere 21: 738–749.

Chapter 3

Basics of Soil Erosion

Manmohanjit Singh[1],* and *Kerstin Hartsch*[2]

Introduction

Soil erosion implies the physical removal of topsoil by various agents, including falling raindrops, water flowing over and through the soil profile, wind velocity and gravitational pull. Erosion is defined as "the wearing away of the land surface by running water, wind, ice or other geological agents, including such processes as gravitational creep". Soil erosion refers to the detachment and carrying away of soil particles to another place by the agencies of water, wind or gravitational forces, etc. Soil erosion is most destructive phenomenon worldwide since it involves not only the loss of water and plant nutrients but ultimately the soil itself. Noticeably or unnoticeably erosion of the soil goes on at all moments and at all places. When the rate of erosion does not exceed the rate of soil formation it is termed as geologic, natural or normal erosion. When the rate of erosion exceeds the rate of soil formation it is called accelerated erosion. The accelerated erosion is primarily man-made because of the effect of agriculture and deforestation.

Accelerated erosion is a serious problem in all climates because wind as well as water can remove soil. It affects both agricultural areas and the natural environment. It has impacts which are both on-site (at the place where the soil is detached) and off-site (Zeneli 2017). The use of powerful agricultural implements

[1] Regional Research Station (Punjab Agricultural University), Ballowal Saunkhri, SBS Nagar, 144521, India.
[2] IPROconsult GmbH, Department of Ecology and Environment, Dresden Germany.
  Email: Kerstin.hartsch@iproconsult.com
* Corresponding author: mmjsingh@pau.edu

has, in some parts of the world, led to damaging amounts of soil moving down slope, under the action of gravity, which is called tillage erosion. It has been estimated that accelerated soil erosion has irreversibly destroyed 30% of the present cultivated area in the world. In general soil erosion is more severe in mountainous and undulating areas.

Causes of soil erosion can be listed as:

- Large scale deforestation.
- Developmental activities, e.g., construction of roads, big dams and mining in regions of very steep slopes.
- Shifting cultivation, wrong agricultural practices and cultivation of fragile areas.
- Over population, harsh climatic conditions, over exploitation and unwise use of soil resources.
- Increased demand for fodder, fuel, timber and additional land.

On-site Effects

Loss of agricultural productivity

Soil erosion's on-site effects are predominant on agricultural lands. It results in loss of soil from the field, redistribution of soil within a field and reduction in soil quality in terms of the breakdown of soil structure, decline in organic matter and nutrients and reduction of cultivable soil depth. The available soil moisture capacity is also reduced resulting in more drought-prone conditions. The net effect is loss of soil fertility and soil productivity, which restricts what can be grown and results in an increased expenditure on fertilizers to maintain yield (Agata et al. 2018).

Economic impact

Increased use of artificial fertilizers may to an extent and for a time, compensate for erosion-induced loss of soil quality, where economic circumstances are favourable. Farmers of developed countries can cope to some extent the loss in soil productivity by applying chemical fertilizers but for the resource poor farmers of rest of the world it is not feasible (Posthumus et al. 2015). These extra costs are necessarily borne by the farmers although they may be passed on in part to the community in terms of higher food prices as yields decline or land goes out of production. At the community level it results in a substantial decline in land value and has consequences for food security.

Other on-site effects such as loss of roads and bridges, forest and grazing lands, loss of animal or human lives by mass movements or landslides, etc., has social and economic effects at the regional, community level or national level.

Off-site Effects

Sedimentation

Sedimentation or silting down streams or downwind is a major off-site problem. It results in reduced capacity of rivers, dams and drainage ditches, enhances the risk of flooding, blocks irrigation canals and shortens the design life of reservoirs. Many hydroelectricity and irrigation projects have been ruined as a consequence of erosion. Improperly designed soil conservation structures or water harvesting structures are also silted up with in a short span of time.

Sedimentation may also be in agricultural fields downstream or downwind, which may leave the land unproductive or unmanageable. Sedimentation results in the pollution of water bodies as sediments may have high quantities of nitrogen, phosphorous and other agro chemicals. These result in eutrophication and the loss of aquatic life.

The indirect impacts of off-site sedimentation on agriculture may be loss of irrigation facilities by the siltation of water harvesting structures and reduced power generation affecting agriculture indirectly. The breakdown of soil aggregates also reduces soil carbon storage as carbon dioxide is released into the atmosphere resulting in global consequences such as climate change and the greenhouse effect. Lal (1995) has estimated that global soil erosion releases about 1.14 Pg C annually to the atmosphere. However, there is an extraordinary variability in soil erosion rates in the world (Garcia–Ruiz et al. 2015).

Flash floods

As water is not retained on the sloping lands due to the absence of natural vegetation, there is very little time for rainwater to infiltrate into the soil. This causes an increase in runoff and flash floods. These floods may cause a loss of property and life. The results of the study conducted by Paix et al. (2011) showed that the use of fuelwood and the competition for agriculture land are the main causes of deforestation, which leads to increased soil erosion and floods.

The economic consequences of off-site effects may be much higher as compared to the on-site effects. The off-site effects are borne by govt. agencies in addition to farmers.

Types of Soil Erosion

Soil Erosion can be classified on various bases:

I. Based on rate of soil erosion

a. Geological erosion

Erosion always takes place naturally. The surface of the earth is constantly changing with mountains rising, valleys being cut deeper and wider, the coastline receding at one place and advancing at another. The physical pattern on the surface of earth is the result of these processes over centuries. Erosion is must for the formation of alluvial soils and sedimentary rocks. In other words when the rate of soil erosion does not exceed soil formation, we call it 'geological erosion' or normal erosion or natural erosion.

b. Accelerated erosion

Due to the activities of man, or when climate or topographic conditions are such that the geological erosion is quicker than usual, it leads to accelerated erosion. In other words when the rate of soil erosion exceeds rate of soil formation, accelerated erosion is said to take place.

The most important activity by man, which results in accelerated erosion, is agricultural activity. Nearly all agricultural operations tend to increase or encourage erosion. When vegetation is cleared the ground is more exposed, and there are fewer trees to slow down the wind, which causes wind erosion. There is also less vegetation to absorb the energy of falling rain, which again results in more soil erosion. By ploughing and tilling soil strength is decreased which may in turn accelerate soil erosion.

II. Based on agents of soil erosion

The main agents that loosen and break down soil particles are water and wind. Other agents like temperature and biological agents are working mostly towards geological erosion.

a. Water

This is the most important single agent of erosion. Rainfall, streams and rivers all scour away or carry away soil. Waves erode the shores of sea and lakes. Water in movement is always eroding at its boundaries.

b. Wind

Wind does not by itself wear away rocks, but abrasion, even of hard rock, resulting from grains of sand or soil carried in suspension cause erosion. Wind erosion takes place normally in arid and semi-arid areas devoid of vegetation, where the wind velocity is high.

c. Temperature

The cracking and flaking of rocks by variations in temperature is a common feature. Rapid variations between day and night temperatures affect the surface of rocks, while the changes due to slower variations between summer and winter penetrate deeper. When the temperature changes include frost, disruption is greatly increased by the expansion of water in cracks and crevices.

d. Biological agents

Living organisms such as lichens and mosses on rocks cause actual destruction. But the main effect of living things is the disturbance, which speeds up the effect of other agents. Animals trampling on rocks or soil break it down and make it more easily carried away by wind or water. Earthworms and termites disturb the soil and increase the aeration and oxidation, and so speed up the process of conversion from resistant rocks to erodible soil.

III. Based on the stages of soil erosion

Water erosion is a two-part process involving the detachment and transport of soil particles. The water erosion process consists of discrete stages from raindrop impact to the formation of gully erosion or mass movement. Each stage has its own processes and characteristics. Controlling or preventing water erosion requires an understanding of each step in the erosion process.

a. Splash erosion (Raindrop impact)

Splash erosion or raindrop impact represents the first stage in the erosion process. Raindrops behave as little bombs when falling on exposed or bare soil, displacing soil particles and destroying soil structure. Studies have shown that splashed particles may rise as high as 0.6 metres above the ground and move up to 1.5 metres horizontally. Splash erosion results in the formation of surface crusts, which reduce infiltration resulting in the start of runoff.

b. Sheet erosion (Inter-rill erosion)

Sheet erosion is defined as the uniform removal of soil in thin layers from sloping land. The top fertile soil layer is washed away from the arable lands. It can be a very effective erosive process because it can cover large areas of sloping land and go unnoticed for quite some time. Sheet erosion can be recognized by either soil deposition at the bottom of a slope or by the presence of light coloured subsoil appearing on the surface. It typically results in the loss of surface soil particles, which contains the bulk of the available nutrients and organic matter. It can also be recognized from the muddy colour of runoff water. Sheet erosion rarely flows for more than a few metres before concentrating into rills, so a better approach is to describe this phase as 'inter-rill' erosion, meaning both the movement by rain splash and transport of raindrop-detached soil by thin-flow surface flow whose erosion capacity is increased by raindrop impact turbulence.

c. Rill erosion

Rill erosion is the most common form of erosion. When sheet erosion is allowed to continue unchecked, the silt-laden runoff forms well-defined small ephemeral channels called rills. The concentrated flow is able to detach and transport soil particles and channels up to 30 cm deep can be formed. The rills formed from one storm are often obliterated before the next storm, when the channels may form an entirely fresh network, unrelated to the position of previous rills. The rill system is discontinuous and has no connection with the main river system. Only occasionally does a master rill develop a permanent cause with an outlet to the river. The rill channels can temporarily be obliterated by tillage. Rill erosion can be prevented by either reducing flow velocity or hardening the soil to erosion.

d. Gully erosion

Gully erosion is an advanced stage of rill erosion where surface channels have eroded to the point where they cannot be removed by tillage operation. Gullies are the most spectacular evidence of the destruction of soil. Gully erosion is responsible for removing vast amounts of soils, irreversibly destroying farmlands, roads and bridges and reducing water quality by increasing the sediment load in streams. Gully erosion occurs when the concentration flow of water along flow routes cause sharp-sided entrenched channels deeper than 0.5 m.

e. Stream bank erosion

Stream bank erosion occurs when streams begin cutting deeper and wider channels as a consequence of increased peak flow or the removal of local protecting vegetation. It results in increase in stream sediment and suspended material. The rivers change their courses due to this type of erosion.

f. Mass movements and land sliding

This type of erosion is in response to the force of gravity accompanied by removal of forests, and construction of hill roads during intense rains. Different forms of mass movements include slumping, soil creep, rotational slip, rock fall, rockslide and mudslide, etc. A landslide is defined as an outward and downward movement of the slope forming material, composed of natural rocks, soil, artificial fills, etc. The fundamental causes of landslides are the topography of the region and geological structure, the kinds of rocks and their physical characteristics. The immediate cause of a slide may be an earthquake or a heavy rainfall.

IV. Specialized forms of water erosion

a. Pedestal erosion

When an easily erodible soil is protected from splash erosion by a stone or tree root, isolated 'pedestals' capped by the resistant material are left standing up from the surrounding ground. The erosion of the surrounding soil is shown to be mainly by splash rather than by surface flow because there is little or no undercutting at the base of the pedestal. This type of erosion develops slowly over several years and is often found on bare patches of grazing land. It helps in deducing approximately what depth of soil loss has been eroded by studying the height of the pedestals.

b. Pinnacle erosion

In highly erodible soils high pinnacles in gully sides or bottoms are found. Deep vertical rills in gully sides cut back rapidly and they join and leave the isolated pinnacle. A more resistant soil layer or gravel or stones, often cap the pinnacle (as in pedestal erosion). This may be due to physical or chemical soil conditions such as excessive sodium and complete de-flocculation.

c. Piping or tunnelling

It occurs when surface water infiltrates through the soil surface, cracks, root channels and animal burrows and moves downwards until it comes to a less permeable layer. If there is an outlet so that the water can flow laterally through the soil over the less permeable layer, then the fine particles of the more porous soil may be washed out. This in turn increases the lateral flow, so the sideways erosion increases, and eventually the whole of the surface flow disappears down a vertical pipe and flows underground probably through the sides of a gully. This tunnelling is an insidious form of sub-surface erosion, resulting in considerable damage even before surface manifestations are evident. Tunnel erosion is particularly difficult and expensive to control and is not always successful. Jakiel and Poesen (2018) reported the significance and research needs of sub-surface erosion by soil piping.

d. Slumping

It is usually a process of geological erosion and although it may be accelerated as with the sides of gullies, it can occur without any intervention of man. It becomes prominent in high rainfall areas with deep soils. In such areas it can become the main agent in the development of gullies. The other main causes of slumping are riverbank collapse and coastal erosion.

e. Fertility erosion

It is the loss of plant nutrients by erosion and can be comparable in magnitude with the removal of the same elements in the harvested crop. Phosphorous is mainly lost along with the colloidal particles on whose surface it is adsorbed. Nitrogen is soluble in the forms of nitrite and nitrate and is lost in solution forms in the runoff without any physical soil movement.

f. Puddle erosion

It is the physical breakdown of soil by rain and washing of fine soil fractures into a depression which results in a structure less soil and choked soil whose productivity decreases.

g. Vertical erosion

It is washing down of fine clay particles through porous sand or gravel to accumulate at some less pervious layer further down the profile. It takes place during puddling soils for rice cultivation.

Factors Affecting Water Erosion

The ecological parameters that influence the effects of the agents (water, wind, etc.) of soil erosion are the factors of soil erosion. The most active factor of soil erosion is humans who can either accelerate soil erosion by misusing land or can curtail it by adopting proper soil and crop management practices. The major variables affecting soil erosion are climate, soil, vegetation and topography. Of these, vegetation and, to some extent, soil and topography may be controlled.

Climate

Climatic factors affecting erosion are precipitation, temperature, wind, humidity and solar radiation. Precipitation is a broad term used for fog, mist, hail, snow and rain. It is rain and snow that play a major role in soil erosion. Temperature and wind are most evident through their effect on evaporation and transpiration; however wind also changes raindrop velocities and the angle of impact. Humidity and solar radiation are somewhat less directly involved in that they are associated with temperature and the rate of soil water depletion.

Rainfall characteristics affecting soil erosion are amount, intensity, distribution, raindrop size, seasonality and variability of rainfall. The most important aspects of rainfall are its total quantity and its intensity. The erosive effect of rain is enhanced by the disaggregating and the splashing effect of raindrops. Besides depending on the disaggregating effect of raindrops, the total amount of eroded soil also depends on the erosive action and transporting capacity of surface flow. Without surface runoff, the amount of soil erosion caused by precipitation is relatively small. Therefore, a critical factor that determines the erosive effect of rainwater is the permeability of the soil, which indirectly influences total soil losses and the pattern of erosion processes on slopes. While the erosive activity (erosivity) of raindrops is determined by the kinetic energy of the raindrops, the erosive action and transporting capacity of surface flow depends on its quantity, velocity and degree of confluence. Rainfall simulation experiments were conducted on two runoff plots with four varying slopes and two rainfall intensities (90 and 120 mm h^{-1}) by Fang et al. 2015. It was observed that higher rainfall intensity produced less runoff and more sediment under all treatments. At lower rainfall intensities a linear function fits the relationship between soil loss and rainfall intensity whereas this function tends to be non-linear at higher intensities. A strong non-linear relationship was found between different quartiles of storms and soil loss (Mohamadi and Kavian 2015).

While the erosive effect of raindrops depends on the size of the soil grains for a given type of soil and on the velocity of the falling raindrops (which is a function of their size), the erosive effect of surface flow depends on the critical velocity of the water and its carrying capacity, which varies according to the soil grains being carried. The erosive action of rain-water increases with increasing size of the raindrops, since larger drops have the effect of reducing soil permeability. As the intensity of rain increases, the contribution made to the overall erosion by surface runoff increases faster than that made by the impact of the rain on the soil. A laboratory study was conducted to quantify the effects of raindrop impact and runoff detachment on soil erosion and soil aggregate loss during hillslope erosion processes (Lu et al. 2016). A soil pan was subjected to different rainfall intensities under two soil surface conditions: with and without raindrop impact by placing a nylon net over the soil pan. The results showed that raindrop impact played the dominant role in hillslope soil erosion and soil aggregate loss. Soil loss caused by raindrop impact was 3.6–19.8 times higher than that caused by runoff detachment. The contributions of raindrop impact to hillslope soil erosion were 78.3% to 95.2%.

Regardless of the relative extents of the various phases of the erosion process, it may be stated that rain intensity is the most important factor governing soil erosion by water. The general rule is that the more permeable the soil, the smaller is the erosive effect of rain, and vice-versa. On comparatively impermeable soils, soil wash occurs provided that the total amount of rainfall is large, even if not of high intensity.

The erosive effect of rain also increases during a succession of downpours. It has been shown that the first rain builds up the soil moisture content, disaggregates the soil clumps by impact or by dissolution, diminishes soil permeability, and to some extent models the mono-relief and the micro-relief. Thus erosion losses increase in successive downpours, although the intensity and amount of rainfall may decrease.

Soil

Although soil resistance to erosion depends in part on topographic position, slope steepness and the amount of disturbance created by man (during tillage) the soil properties are the most important determinants. The corresponding soil characteristics that describe the ease with which soil particles eroded are soil detachability and soil transportability, which when combined called soil erodibility. Soil erodibility, which is the resistance of the soil to both detachment and transport, depends on the physical and hydrological, chemical and mineralogical, and biological and biochemical properties as well as soil profile characteristics. Important soil physical properties that affect the resistance of a soil to erosion include texture, structure, water retention and transmission properties and shear strength.

Soil with a sufficiently high permeability to absorb precipitation of maximum intensity (about 5 mm min^{-1}) is only seldom affected by sheet erosion, and is therefore, damaged only by splash erosion. However, soils of this type show only little resistance to erosion and any confluence of surface water easily carves rills, which attain a considerable size during heavy rains.

Low permeability soil and impermeable soils, on the other hand, have more resistance, but then much greater surface runoff develops on soils of this type. Erosion increases if soil permeability is reduced artificially, or if surface layers are loosened as a result of soil cultivation. Soil permeability, as well as resistance to erosion may be increased by improving the soil structure, especially if the proportion of water table aggregates is increased.

In addition to other soil properties, soil erodibility can be determined by soil texture, active surface area of particles and by the homogeneity of granulation, etc. The coarser the soil texture, the smaller the active surface area and the more homogeneous the granulation, the smaller the resistance of the soil to erosion. Since all these properties are altered by the selective action of erosion and by the transport of the particles loosened by erosion, soil that has already been transported is less resistant to erosion.

Wischmeier and Smith (1978) assessed soil erodibility in terms of the depth of removed soil divided by the index of rainfall-mediated erosion. These parameters were based on measurements made on denuded soil on experimental plots of constant dimensions (22 m length, 9% slope inclination). Soil erodibility assessed in this way is doubtless the nearest to reality, yet it depends on the dimensions of the experimental plots, and, to a certain extent is distorted by the amount of surface runoff. Wischmeier et al. (1971) developed a nomograph to estimate soil erodibility but that nomograph was not applicable in many soil conditions and was modified (Singh and Khera 2009, Aureswald et al. 2014).

Relief and slope steepness

Topographic features that influence erosion are degree of slope, shape and length of the slope, and size and shape of the watershed. Erosion would normally be expected to increase with increase in slope steepness and slope length as a result of respective increase in velocity and volume of surface runoff. On longer slopes, an increased accumulation of overland flow tends to increase rill erosion. Concave slopes, with lower slopes at the foot of the hill, are less erosive than convex slopes. The relief of the terrain is of fundamental importance in determining levels of water erosion. Included among the factor of relief are: slope inclination and length, slope form, modeling of the relief, slope aspect, and affecting erosion indirectly-elevation above sea level.

As the slope becomes steeper, the runoff coefficient increases, the kinetic energy and carrying capacity of surface flow become greater, soil stability and slope stability decreases, splashing erosion increases, and the possibility of soil displacement in a downhill direction during ploughing is greater. Thus the likelihood of soil erosion increases with the growing steepness of the slope.

Slope length

Slope length is important mainly with respect to the increase in the flow of water on slopes and the degree of confluence. As the quantity of water and its degree of confluence grow, the velocity and transporting capacity change. In general, with the growing length of the slope the multiple of erosion intensity decreases, although the absolute differences have an increasing tendency.

Slope aspect

The effect of slope aspect operates through the different degrees of isolation occurring on sunny versus shaded slopes. With the higher temperatures attained on sunny slopes, the rate of decomposition of organic matter, the rate of evapotranspiration, the degree of salt concentration, and other processes all increase. The aspect affects desiccated soil mainly.

Elevation above the sea level and geographical location

Both these factors have an indirect effect on erosion by their influence on physical conditions. With increasing elevation and at higher geographical latitudes, the temperature generally decreases and the amount of precipitation increases. However, the intensity of precipitation increases with elevation and decreases at higher latitudes. The combined influence of temperature, precipitation, wind, potential energy of the relief, and surface structure affect the erosive activity of both precipitation and wind. These, together with other geomorphic factors create a varied pattern of destruction phenomena.

Vegetation

Vegetation acts as a protective layer or buffer between the atmosphere and the soil. The major effects of vegetation in reducing erosion are:

a. Interception of rainfall by absorbing the energy of the raindrops and thus reducing surface sealing and runoff.
b. Retardation of erosion by decreased surface velocity.
c. Physical restraint of soil movement.
d. Improvement of aggregation and porosity of the soil by roots and plant residues.
e. Increased biological activity in the soil.
f. Transpiration, which decreases soil water, resulting in increased storage capacity and less runoff.

These vegetative influences vary with the season, crop, degree of maturity of the vegetation, soil and climate, as well as, with the kind of vegetative material, mainly roots, plant tops and plant residues.

Agricultural measures

Soil cultivation plays an important part in the reduction of erosion mainly on account of the effect on surface roughness, soil permeability, soil resistance against destruction caused by raindrops and surface runoff freezing of the soil, and the mobilization of nutrients and water for plant growth. It is generally accepted that soil cultivation, fertilizing, irrigation and crop distribution according to rotation practice are basic soil conservation measures applied on agricultural land by means of which erosion on land of low and medium erodibility may be reduced to a harmless level. In addition, also mulching of the ground and its reinforcement by incrustation are important factors in the reduction of erosion. Mulching practices have been globally found to be effective in controlling soil erosion (Prosdocimi et al. 2016).

Soil moisture content

Increase in moisture content of a soil decreases its shear strength and bring about changes in its behaviour. At low moisture contents the soil behaves as a solid and fractures under stress but with increasing moisture content it becomes plastic and yields to flow without fracture. The point of change in behaviour is termed the plastic limit. With further wetting, the soil will reach its liquid limit and start to flow under its own weight. The instantaneous soil moisture content plays a vital role during successive rainstorms or during downpours, which occur, in the dry season. It has been generally established that the higher the soil moisture, the lower the resistance of the soil to erosion, these properties being associated both with infiltration and with the resistance of oil aggregates.

Infiltration capacity

The maximum sustained rate at which soil can absorb water is influenced by pore size, pore stability and the form of the soil profile. Soils with stable aggregates maintain their pore spaces better while soils with swelling clays or minerals that are unstable in water tend to have low infiltration capacities. In layered soils, it is the layer with the lowest infiltration capacity, which is critical. In sandy soils, formation of crust results in decreased infiltration. Increasing intensity of rain may not lead to a corresponding increase in runoff and decreasing intensity may even lead to runoff.

Rainfall erosivity

Rainfall erosion is the interaction of two factors—the rain and the soil. The amount of erosion, which occurs in any given circumstances, will be influenced by both. It has been established that one storm can cause more erosion than another on the same land and the same storm causes more erosion on one field than on another. This effect of rain is called erosivity and the effect of the soil is called erodibility. Erosivity is the potential ability of rain to cause erosion. It is a function of the physical characteristics of rainfall. Erodibility on the other hand is the vulnerability or susceptibility of the soil to erosion and it is a function of both the physical characteristics of the soil and the management of the soil. A value on the scale of erosivity depends solely on rainfall properties, and to this extent it is independent of the soil. But a quantitative measurement of erosivity may only be made when erosion occurs, and this involves the erodibility of the eroded material.

Similarly the relative values of erodibility are not influenced by rain, but can only be measured when caused by rain, which must have erosivity. Thus neither is independently quantitative but may be studied quantitatively while the other is held constant.

Rainfall erosivity

Soil erosion is a work process in the physical sense that work is the expenditure of energy, and energy is used in all the phases of erosion—in breaking down soil aggregates, in splashing them in the air, in causing turbulence in surface runoff, in scouring and carrying away soil particles. In Table 1, the kinetic energy available from falling rain is compared with that from surface runoff. The exact figures used in this calculation are not important since they are based upon assumptions of the percentage runoff and assumed velocities but clearly the difference in the amount of energy is very large, with rainfall energy dominating the picture.

The rain thus has 256 times more kinetic energy than the surface runoff. The principal effect of raindrops is to detach soil, while that of surface flow is the transportation of the detached soil.

Raindrop impact has other important effects as well as particle detachment. The detached particles lead to the sealing of the soil surface and hence to lower infiltration and increased surface runoff. The rain energy causes turbulence in the runoff, thus greatly increasing its capacity to scour and to transport soil particles.

Table 1. Kinetic energy of rain and runoff.

Character	Rain	Runoff
Mass	Assuming the mass of falling Rain is R	Assuming 25% runoff mass of runoff is R/4
Velocity	Assume terminal velocity of 8 m/s	Assume speed of surface flow of 1 m/s
Kinetic energy	$\frac{1}{2} * R(8)^2 = 32\,R$	$\frac{1}{2}.R/4.\,(1)^2 = R/8$

Rainfall erosivity indices

The most suitable expression of the erosivity of rainfall is an index based on the kinetic energy of the rain. Thus the erosivity of a rainstorm is a function of its intensity and duration, and of the mass, diameter and velocity of the raindrops. To compute erosivity requires an analysis of the drop-size distributions of rain. It has been shown that drop-size characteristics vary with the intensity of rain; for example, the median drop diameter (D_{50}) increases with rain intensity. However, in tropical countries it has been shown that this relationship holds only for rain intensities up to 100 mm h^{-1}. At greater intensities, D_{50} decreases with increasing intensity, presumably because greater turbulence makes larger drop sizes unstable.

To compute the kinetic energy of a storm, a trace of the rainfall from an automatically recording rain gauge is analyzed and the storm divided into small time increments of uniform intensity. For each time period, knowing the intensity of the rain, the kinetic energy of rain at that intensity is estimated using the equation:

$$K.E. = 11.87 + 8.73 \log_{10}I$$

This K.E. multiplied by the amount of rain gives the kinetic energy for that time period. The sum of the kinetic energy values for all the time periods gives the total kinetic energy of the storm. To be valid as an index of potential erosion, an index must be significantly correlated with soil loss.

EI_{30}: Studies have shown that soil loss by splash, overland flow and rill erosion is related to a compound index of kinetic energy and the maximum 30-minute rainfall intensity (I_{30}). T is the greatest average intensity experienced in any 30-minute period during a storm. It is computed from recording rain gauge charts by locating the greatest amount of rain which falls in any 30 minutes, and then doubling this amount to get the intensity (rainfall per hour). It can be computed for individual storms, and the storm values can be summed up over periods of time to give weekly, monthly or annual value of erosivity.

This EI index is being criticized for the following reasons. Firstly it is based on estimates of kinetic energy and using the empirical equation is not valid for tropical rains of high intensity. Secondly, it assumes that erosion occurs even with light intensity rain. The inclusion of I_{30} in the index is an attempt to correct for overestimating the importance of light intensity rains but it is not entirely successful. In fact, there is no obvious reason why the maximum 30-minute intensity is the most appropriate parameter to choose. At some places with sparse and dense plant cover 15 and 5 minute intensities have shown better results.

KE > 25 index: It is based on the fact that little erosion takes place at low intensities. At low intensity, rain is composed mainly of small drops, falling with low velocity, and hence low energy. Even if a little splash erosion occurs, there is usually no runoff to carry away the splashed particles. Studies have shown that although there is variation from storm to storm, the intensity of 25 mm per hour can be taken as a threshold value separating erosive and non-erosive rain. KE > 25 index means summing the kinetic energy received in those time increments when the rainfall intensity equals 25 mmh^{-1} or greater. The index has been modified for temperate regions using a lower threshold value of 10 mmh^{-1}.

AI_m: It is the product of amount of rain (A) and maximum intensity over a 7.5 minute period. It correlated best with soil loss from small plots in Nigeria.

Stages of Soil Erosion

Rain splash erosion

Falling raindrops are the major agents responsible for initiating soil erosion, i.e., causing soil detachment and displacement from its original position, although the impact of raindrops of shallow streams may not splash soil; it does increase turbulence, providing a greater sediment carrying capacity. Inter-rill erosion occurs on an area where all detachment is due to the forces of raindrop impact and transport is primarily by overland flow. Tremendous quantities of soil are splashed into the air, most particles more than once. The amount of soil splashed into the air as indicated by the splash losses from small elevated pans, was found to be 50 to 90 times greater than the runoff losses. On base soil it is estimated that as much as 200 Mgha^{-1} is splashed into the air by heavy rains. Single drop studies have examined the shapes of raindrops, forces of raindrop impact and reaction of soil to the impact forces.

Splash erosion is affected by raindrop mass, size distribution, shape, velocity and direction. Drop shape affects the amount of soil splash. By changing the height of fall from 0.57 to 0.62 to 6.67 m, the amount of splash loss changed from 0.78 to 0.28 to 0.88 g per drop. The drop shape at 0.57 and 0.67 m fall height was oblate.

The action of raindrops on soil particles is most easily understood by considering the momentum of a single raindrop falling on a sloping surface. The down slope component of this momentum is transferred in full to the soil surface but only a small proportion of the component normal to the surface is transferred, the remainder being reflected. The transfer of momentum to the soil particles has two effects. First, it provides a consolidating force, compacts the soil; second, it produces a disruptive force as the water rapidly disperses from and returns to the point of impact in laterally flowing jets. Raindrops due to friction with soil surfaces are broken up into finer droplets which have local velocities double than that of the original raindrops and are sufficient to impart a velocity to some of the soil particles, launching them into the air. Thus raindrops are both agents of consolidation and dispersion.

The consolidation effect is best seen in the formation of a surface crust, usually only a few mm thick, which results from the clogging of the pores by compaction. This is associated with the dispersal of five particles, from soil aggregates or clods which are translocated to infill the pores. The most important effect of a surface crust is to reduce infiltration capacity and thereby promote greater surface runoff. Crustability decreases with increasing contents of clay and organic matter since these provide greater strength to the soil. The detailed information on soil crusting has been shared elsewhere in the book.

Soil Splash is influenced by many factors including antecedent soil properties, landform, rainfall characteristics, properties of overland flow and vegetation cover (Table 2). When raindrops hit dry soil aggregate, the energy of the raindrop is transmitted to the soil aggregate. The aggregate gets wet, its soil moisture potential increases, its strength decreases and its particles are detached; later on the entrapped air can virtually explode, breaking the aggregate and spattering soil particles into the air. The heat of wetting or the energy released when soil water potential changes from one energy state to another, plays a significant role in the detachment of relatively dry soil. Fast release of heat of wetting causes more soil detachment and splash.

Minimal energy is needed for soils with a geometric mean particle size of 0.125 mm and the soils with geometric mean particle size between 0.063 and 0.250 mm are the most vulnerable to detachment. Coarser soils are resistant to detachment because of the weight of the large particles. Finer soils are resistant because the raindrop energy has to overcome the adhesive or chemical bonding forces that link the minerals comprising the clay particles. Overall, silt looms, loams, fine sands and sandy loams are the most detachable. Selective removal of particles by rain splash can cause variations in soil texture down slope. Splash erosion was observed to be higher in sandy loam than in silt loam as measured under both the drop size spectra of 2.5 and 3.5 mm (Kukal and Sarkar 2011). The raindrop size spectra of 3.5 mm could easily breakdown even the strongest aggregates of silt loam, resulting in higher splash erosion ($343.5 \ g \ m^{-2}$) than under 2.5 mm drop size ($114.2 \ g \ m^{-2}$). Surface compaction increased average splash loss (40.5%) under drop size spectrum of 3.5 mm. The average splash erosion decreased by 50% with bigger aggregates at soil surface (1–2 mm) from that with smaller aggregates (0.5–1 mm). With further increase in aggregate size from 1–2 mm to 2–4 mm, the splash erosion decreased by 20.5%.

Soil splash is related to rainfall amount and intensity. For a given amount of rainfall, high intensity rain produces more splash than rain at low intensity. Many site-specific empirical regression equations relating soil splash to rainfall intensity have been developed. Soil splash depends on the kinetic energy of the impacting raindrop, and hence on its size. The size of the soil particle displaced depends on

Table 2. Factors affecting soil splash.

Soil Properties	Soil moisture potential, particle size distribution, soil structure, organic matter, bulk density, exchangeable cations and shear strength.
Landforms	Slope steepness, slope shape and aspect, slope length.
Rainfall characteristics	Mass, size, shape and impact velocity of raindrop, kinetic energy and momentum, intensity, wind velocity
Overland flow	Depth, type of flow, i.e., laminar or turbulent
Vegetation caver	Canopy cover, foliage distribution.

the terminal velocity of the impacting drops. There is a threshold impact velocity below which the soil particles are not displaced by raindrop impact. Fu et al. (2016) derived a relation between the amount of splash detachment, drop size and distance of splash detachment. The relationship is as follows:

$$M = 0.741D^{4.846} \times S^{-1.820}$$

where M is the splash detachment (grams), D is the drop size (mm) and S is splash detachment (cm).

Rain does not always fall on to a dry surface. During a storm it may fall on the surface water in the form of puddles or overland flow. The ability of a raindrop to cause detachment and soil splash differs when overland flow is present or absent. As the thickness of the surface water layer increases, so does splash erosion. It is because of the turbulence that the impacting raindrops impart to the water. There is however a critical water depth beyond which erosion decreases exponentially with increasing water depth because more of the rainfall energy is dissipated in the water and does not affect the soil surface. In addition to soil splash, raindrop impact in the overland flow may increase the transport capacity of the overland flow. There is movement of even bigger sized particles when they are submerged in water.

Wind speed impacts a horizontal force to a falling raindrop until its horizontal velocity component equals the velocity of the wind. As a result, the kinetic energy of the raindrop is increased. Detachment of soil particles by impacting wind driven raindrops can be some 1.5–3.0 times greater than that resulting from rains of the same intensity without wind.

Vegetation cover may dissipate raindrop impact and protect the soil against splash. Vegetation cover alters the volume, drop size distribution, impact velocity and kinetic energy of the rainfall reaching the ground. The effect of vegetation cover as splash erosion depends on many factors like, foliage characteristics, canopy height and ground cover percentage. Canopy cover may not be effective in controlling splash erosion particularly when the canopy height is more. The reason may be that when coalesced drops fall from large trees, they often reach the terminal velocity and have high kinetic energy.

Since splash erosion acts uniformly over the land surface its effects are seen only where stone or tree roots selectively protect the underlying soil and splash pedestals or soil pillars are formed. Such features frequently indicate the severity of erosion. If raindrops fall on crop residue or growing plants, the energy is absorbed and thus soil splash is reduced. Raindrop impact on bare soil not only causes splash but also decreases aggregation and causes deterioration to soil structure. The most important contribution of splash erosion is to deliver detached particles to overland flow, which was the main agent of sediment transport in the inter-rill areas.

Overland flow

Overland flow is an important agent of water erosion. Overland flow is water that flows over the land surface en route to stream channels. It is the initial phase of surface runoff that eventually becomes a major agent of sediment detachment, entrainment and deposition. Although overland flow is visualized as a broad sheet flow, it includes many shallow but easily definable channels.

Overland flow occurs on hillsides during a rainstorm when:

- Surface depression storage is exceeded.
- In the case of prolonged rain, soil moisture storage is exceeded.
- With intense rains, the infiltration capacity of soil is exceeded.

Excess rainfall over infiltration in first used to fill all the depression storages, which may range from 2.5 cm for smooth-surface clay to 5.0 cm for sandy soils. Depression storage may be far greater in soil with stubble and vegetation cover than in bare soil.

Shallow flow may be laminar, turbulent or both. Areas of turbulent flow are often interspersed with areas of laminar flow. Turbulence is caused by following raindrops and wind driven rain. Turbulent flow is the most relevant to the soil erosion. The hydraulic characteristics of the flow are described by its Reynolds number (Re) and its Froude number (F), defined as follows:

$$R_e = \frac{Vr}{\upsilon}, F = \frac{V}{\sqrt{gr}}$$

where V is velocity of water, υ is kinematic viscosity, r is hydraulic radius, which for overland flow, is taken as equal to flow depth.

The Reynolds number is an index of the turbulence of the flow. The greater the turbulence, the greater is the erosive power generated by the flow. At numbers less than 500, laminar flow prevails and at values above 2000, flow is fully turbulent. In turbulent flow, the water moves in highly irregular paths, causing an exchange of momentum from one portion of water to another. The turbulence increases shear stresses throughout the fluid. In laminar flow, each fluid layer moves in a straight line with uniform velocity and there is no mixing between the layers. Intermediate values are indicative of transitional or disturbed flow, often a result of turbulence being imparted to laminar flow by raindrop impact.

The Froude number is an index of whether or not gravity waves will form in the flow. When the Froude number is less than 1.0, gravity waves do not form and the flow, being relatively smooth, is described as tranquil or sub critical. Froude number greater than 1.0 denote rapid or supercritical flow, characterized by gravity waves, which in more erosive. Most overland flow are super critical and Froude no's can be as high as 15. Steady flow occurs when conditions (velocity, density, presence and temperature at any point in water) do not change with time. The flow is unsteady when conditions at any point change with time. Uniform flow occurs when the velocity vector at every point is identical (in magnitude and direction) for any given instant. In non-uniform flow, velocity vector varies from place to place at any given instant.

Rill erosion

Rills are usually described as small, intermittent water courses that present no obstacles or impediments to tillage operations using conventional equipment. Rill erosion is the predominant form of erosion under most conditions. Rills also carry the connotation that once obliterated, the will not inherently reform at precisely the same location. Even excessive inter rill erosion can go unnoticed, but rill erosion is easily observed. Rills are initiated at a critical distance down slope, where overland flow becomes channelled. The depth and velocity of water in channelled flow are much greater than in pre-channel flow. The depth of channelled flow may be 50 times that of overland flow, and the velocity 10 times greater. Shear stress exerted by the concentrated flow causes soil detachment along channel sides and floor. Studies of the hydraulic characteristic of the flow show that the change from overland flow to rill flow passes through four stages, i.e., un-concentrated overland flow, overland flow with concentrated flow paths, micro channels without head cuts and micro channels with head cuts (Wang et al. 2014). At the point of rill initiation, flow conditions change from sub critical to supercritical.

Both depth and velocity of flow are important in determining rill erosion. There is also a critical value of shear velocity before rill erosion begins which is normally 3.0–3.5 cm S^{-1}. The critical shear velocity of rill initiation (U_{crit}) is linearly related to the shear strength of the soil (T_s).

$$U_{crit} = 0.9 + 0.3T_s$$

The shear strength or erosion power of water is increased by sediment concentration. For an equal volume, sediments have 2.65 timer more energy than water and thus highly abrasive. Yao et al. (2008) and Jiang et al. (2018) found that slope was relatively more important than rainfall intensity in determining the location of rill initiation. Soil critical shear stress determined in this study ranged from 1.33 to 2.63 Pa, with an average of 1.94 Pa. Soil critical shear stress was inversely related to slopes and was not influenced by rainfall intensity.

Rill Erosion is a function of the hydraulic shear T of the water flowing in the rill and two soil properties the rill erodibility Kr and the critical shear Tc, the shear below which soil detachment is negligible. Detachment rate Dr is the erosion rate occurring beneath the submerged area of the rill.

$$D_r = K_r(T - T_c)\left(1 - \frac{Q_s}{T_c}\right)$$

where Qs is the rate of sediment flow in the rill and Tc is sediment transport capacity of rill. Once rills have been formed their migration upslope occurs by the retreat of head cut at the top of the channel. The rate of retreat is controlled by the cohesiveness of the soil, the height and angle of the head wall, the discharge and the velocity of the flow. Mass failure of the side walls can contribute more than half of the sediment removed in rills, particularly when heavy rains follow a long dry period during which cracks have developed in the soil.

Rill erosion may account for the bulk of the sediment removed from a hillside, depending on the spacing of the rill and the extent of the area affected. The inter rill and rill erosion processes are used in several process-based erosion prediction computer models.

Gully erosion

Gully erosion is a highly visible form of soil erosion that affects soil productivity, restricts land use and can threaten roads, fences and buildings. The large channels that cannot be removed by tillage are called gullies. A large gully is also called a ravine. Gullies are relatively steep sided watercourses which experience ephemeral flows during heavy or extended rainfall. A gully channel may be U or V shaped depending upon the strength of the sub soils' resistance or its resistance to water's cutting action. Gullies are formed when the surface and sub soil materials are uniformly weak. V shaped gullies are formed when the sub soil is more resistant to erosion than the surface soil. Gullies are having relatively greater death and smaller width, carry large sediment loads and display very erratic behaviour so that relationships between sediment discharge and run off are frequently poor. Gully erosion has become a field of growing interest among the research community but there sre still are numerous knowledge gaps that need to be addressed (Castillo and Gomez 2016).

Gullies are almost always associated with accelerated erosion and therefore with landscape instability. Ephemeral gullies are the channels intermediate in size between rills and classical gullies. They are larger than rills but are small enough to be obliterated by usual farming practices. This gullies reform at same location year by year. Processes involved in transition from rill to gully erosion are not well understood. As a simple guide it is taken that when the cross sectional area of channels increases than 1 m^2, these are called gullies.

Gully erosion is caused when run off concentrates and flows at a velocity sufficient to detach and transport soil particles. In cultivated area or in pasture, advanced rill erosion can develop into gully erosion if no protective measures are taken. Cattle beds can be a starting point for a small rill that can develop into a large gully.

In the first stage of gully formation, small depressions or knicks form on a hill side as a result of localized weakening of vegetation cover by grazing or by fire. Water concentrates in these depressions and enlarges them until several depressions coalesce and an incipient channel is formed. Erosion is concentrated at the heads of the depressions where near vertical scraps develop over which superficial flow occurs. Some soil particles are detached from the scrap itself but most erosion is associated with scouring at the base of the scrap which results in deepening of the channel and undermining of the headwall, leading to collapse and retreat of the scrap upslope. Sediment is also produced further down the gully by stream bank erosion. This occurs partly by the scouring action of the running water and the sediment it contains, and partly by the slumping of banks following saturation during flow.

Widening of the gully sides may occur by slumping and man movement especially on the outside curve of meanders. Scouring of the toe slope can lead to mass failure of the side of the gully under gravity. This soil is then washed away by subsequent flow. Gullies are also formed by piping from subsurface runoff. Tunnelling is an important mechanism for headword and lateral gully expansion in dispersible soils. Cracks develop into tunnels when water flows through them and they soon collapse causing rapid progression of the gully head.

Gully depth is often limited by the depth to the underlying rock which means that gullies are normally less than 2 m deep. However on deep alluvial and colluvial soils, gullies may reach depths of 10 to 15 m. The rate of gully erosion depends primarily on the run off producing characteristic of the watershed, the drainage area, soil characteristics, the alignment, size and shape of the gully and the slope of the channel. Gullies pass through successive cycles of erosion and deposition. It is not uncommon for the head of a gully to be extremely active while the lower section of the gully is stabilizing.

Gully development may be triggered by

- Cultivation or grazing of soils susceptible to gully erosion.
- Increased run off from land use changes such as deforestation in catchments or construction of new residential areas.
- Run off concentration caused by furrows, contour banks water ways, dam by-washes, fences, tracks or roads.
- Improper design, construction or maintenance of waterways in cropping areas.
- Poor vegetative cover, e.g., from overgrazing, fire or salinity problem.
- Low flows or seepage flows over a longer period
- Diversion of a drainage line to an area of high risk to erosion.

Mass movements

When there is an instantaneous movement of a large volume of soil mass or rock material down the slope, it is called mass soil movement. Although mass movement has been widely studied by geologists, geomorphologists and engineers, it is generally neglected in the context of soil erosion.

The presence of moisture in the soil mass adds weight to it which destroys the cohesive properties of the soil. Generally the movement of soil mass is triggered by the action of gravity. The factors responsible for soil mass movement includes the natural or artificial modification to the gradient of the slope, the erosion of the bottom line of a sloping land surface owing to the water flow in a channel or a water courses and the mining activities down the slope, etc. Changes in the water content of rocks owing to precipitation, seepage of water through cracks in clayey layers, ground water level fluctuations, frost and changer in earth's temperature, rock weathering and changes in vegetation cover, etc. have a profound effect on changing the cohesive properties of soil and equilibrium of ground forces. Earthquake and very heavy cloudbursts and presence of rocks with preferential fracture planes can trigger mass movements. Human beings can cause more such mass movements by altering the external geometry of a slope (by terracing, cutting into it to build roads or houses, overloading it with landfills, altering natural flows, etc.).

The stability of the soil mass on a hill slope in respect of mass movement can be assessed by a safety factor (F), defined as the ratio between the total shear strength (σ) of the soil material along a given surface and the amount of stress (T) developed along the surface. Thus

$$F = \frac{\sigma}{T}$$

The slope is stable if F > 1 and failure occur if F < 1.

The many forms of soil mass movement can be broadly divided into two categories, i.e., slow movements and rapid sliding.

Slow movement

Creep: This is a relatively slow sliding of the surface layers of the soil cover, generally without detachment and is widely observed on steep slopes where young fresh samplings are bent and the base of adult trees crooked.

Mechanical or dry erosion: It is caused by cropping techniques. There is steady downward movement of earth pushed by tillage implements. The process eventually scours the hill tops and clog slope bottoms.

Rapid sliding

Debris flow is a generic term used to describe the rapid movement of rocks, soil, water and vegetation downhill. A debris flow could be a mudslide or a landslide, depending on the amount of water present. Flow contains many different size particles from sand grains to boulders, but the bigger rocks travel at the front of the flow.

Landslides: Landslides are at the drier end of the debris flow spectrum. Landslides come in two forms, i.e., block slips and rotational block slips.

Mudslides: Mudslides contain more water than landslides. They can contain solid material, too, but generally have fewer large rocks and trees than landslides.

Local forms: This category includes rock slides, the undermining of banks and slope subsidence leading to localized sliding. These are very frequent at gully heads.

References

Agata, N., A.M. Pisciotta, A. Minacapilli, F. Maltese, A. Capodici, A. Cerda and L. Gristina. 2018. The impact of soil erosion on soil fertility and vine vigor. A multidisciplinary approach based on field, laboratory and remote sensing approaches. Sci. Total Environ. 622-623: 474–480.

Auerswald, K., P. Fiener, W. Martin and D. Elhaus. 2014. Use and misuse of the K factor equation in soil erosion modeling: An alternative equation for determining USLE nomograph soil erodibility values. Catena. 118: 220–225.

Castillo, C. and J.A. Gomez. 2016. A century of gully erosion research: Urgency, complexity and study approaches. Earth-Sci. Rev. 160: 300–319.

Fang, H., L. Sun and Z. Tang. 2015. Effects of rainfall and slope on runoff, soil erosion and rill development: an experimental study using two loess soils. Hydrol. Process. 29: 2649–2658.

Fu, Y., G. Li., T.Z.B. Li and T. Zhang. 2016. Impact of raindrop characteristics on the selective detachment and transport of aggregate fragments in the Loess Plateau of China. Soil Sci. Soc. Am. J. 80: 1071–1077.

Garcia-Ruiz, J.M., S. Begueria, E. Nadal-Romero, J.C. Gonzalez-Hidalgo, N. Lana-Renault and Y. Sanjuan. 2015. A meta-analysis of soil erosion rates across the world. Geomorphology. 239: 160–173.

Jakiel, A.B. and J. Poesen. 2018. Subsurface erosion by soil piping: significance and research needs. Earth-Sci. Rev. 185: 1107–1128.

Jiang, F., Z. Zhenzhi, J. Chen, J. Li, M. Kuang, W. Hongli and G.Y. Huang. 2018. Rill erosion processes on a steep colluvial deposit slope under heavy rainfall in flume experiments with artificial rain. Catena. 169: 46–58.

Kukal, S.S. and M. Sarkar. 2011. Laboratory simulation studies on splash erosion and crusting in relation to surface roughness and raindrop size. J. Indian Soc. Soil Sci. 59: 87–93.

Lal, R. 1995. Global soil erosion by water and carbon dynamics. *In*: Lal, R., J.M. Kimble, E. Levine and B.A. Stewart (eds.). Soils and Global Change. CRC/Lewis Boca Raton, FL: 131–141.

Lu, J., F. Zheng, G. Li, F. Bian and J. An. 2016. The effects of raindrop impact and runoff detachment on hillslope soil erosion and soil aggregate loss in the Mollisol region of Northeast China. Soil Till. Res. 161: 79–85.

Mohamadi, M.A. and A. Kavian. 2015. Effects of rainfall patterns on runoff and soil erosion in field plots. Int. Soil Water Conserv. Res. 3: 273–281.

Paix, M.J., L. Lanhai, C. Xi, S. Ahmed and A. Varenyam. 2011. Soil degradation and altered flood risk as a consequence of deforestation. Land Degrad. Dev. 24: 478–485.

Posthumus, H., L.K. Deeks, R.J. Rickson and J.N. Quinton. 2015. Costs and benefits of erosion control measures in the UK. Soil Use Manage. J. 31: 16–33.

Prosdocimi, M., P. Tarolli and A. Cerdà. 2016. Mulching practices for reducing soil water erosion: A review. Earth-Sci. Rev. 161: 191–203.

Singh, M.J. and K.L. Khera. 2009. Nomographic estimation and evaluation of soil erodibility under simulated and natural rainfall conditions. Land Degrad. Dev. 20: 471–480.

Wang, B., G.H. Zhang, Y.Y. Shi and X.C. Zhang. 2014. Soil detachment by overland flow under different vegetation restoration models in the Loess Plateau of China. Catena. 116: 51–59.

Wischmeier, W.H., C.B. Johnson and B.V. Cross 1971. A soil erodibility nomograph for farmland and construction sites. J. Soil Water Cons. 26: 189–92.

Wischmeier, W.H. and D.D. Smith. 1978. Predicting rainfall erosion losses.USDA Agricultural Service Handbook 537.

Yao, C., T. Lei, W.J. Elliot, D.K. McCool, J. Zhao and S. Chen. 2008. Critical conditions for rill initiation. Trans. Am. Soc. Agric. Biol. Eng. 51: 107–114.

Zeneli, G., S. Loca, A. Diku and A. Lila. 2017. On-Site and Off-Site Effects of Land Degradation in Albania. Ecopersia. 2017: 1787–1797.

Chapter 4

Measurement of Soil Erosion by Water

Manmohanjit Singh[1],* and SS Kukal[2]

Introduction

Accelerated soil erosion as a serious global problem is widely recognized; therefore the assessment of soil erosion is of utmost importance. It is difficult to assess reliably and precisely the extent, magnitude and rate of soil erosion and its economic and environmental consequences. Information readily available in the relavant literature is often based on reconnaissance surveys and extrapolations based on sketchy data. At present the quality of available data is extremely uneven. Land use planning based on unreliable data may lead to costly and gross errors. Standardization of erosion hazard assessments and measurement of different types or processes of erosion is important for the adoption of proper soil conservation measures and land use policy.

Soil Erosion Hazard Assessment

The aim of soil erosion hazard assessment is to identify those areas of land where the maximum sustained productivity from a given land use is threatened by

---
[1] Regional Research Station (Punjab Agricultural University), Ballowal Saunkhri, SBS Nagar, 144521, India.
[2] College of Agriculture, Punjab Agricultural University, Ludhiana, 141004, India.
  Email: sskukal@rediffmail.com
* Corresponding author: mmjsingh@pau.edu

excessive soil loss or the off-site damage arising from erosion is unacceptable. Potential erosion risk takes into consideration the local condition of soil, climate and slope, whereas, the actual erosion risk is greatly modified by the land cover. Therefore, based on soil, climate and slope the area can be designated as having high risk but because of vegetation, it may actually be having a low erosion risk. Assessment of soil erosion can be done by taking into consideration the rainfall, soil and slope data which can be obtained from soil surveys conducted at regional or national levels. Other way is to conduct a detailed soil survey at the field scale at many locations and then extrapolate the information at a national or regional level.

Generalized assessment

The generalized assessment is either based on rainfall data or some indices, such as factorial scoring are also used for this purpose.

Based on rainfall data

Erosivity data can be used as an indicator of regional variation in erosion potential. The mean annual erosivity values can be used to classify areas according to erosion risk. Temporal variation in erosion risk can be assessed using mean monthly erosivity values. Maps of erosivity using the rainfall erosion index R have been produced for the USA (Wischmeier and Smith 1978) and other parts of the world (Panagos et al. 2014). In many countries insufficient rainfall records from autographic gauges are available to calculate erosivity nationwide. In such cases more widely available rainfall parameters, which significantly correlate with erosivity and from which erosivity values might be predicted using a best-fit regression equation, can be used. But care should be there not to extrapolate these results to other locations.

Rainfall intensity data may not be available at many locations. Another parameter to overcome it is to use rainfall aggressiveness. The most commonly used index of rainfall aggressiveness, which is significantly correlated with sediment yields in rivers (Fournier 1960), is the ratio of p^2/P, where p is the highest mean monthly precipitation and P is the mean annual precipitation. Its high value denotes a strongly seasonal climatic regime. Morgan (1976) used data from 680 rainfall stations and found a low but significant correlation between p^2/P and drainage texture. As drainage texture represents gully density so p^2/P may be regarded as an indicator of the risk of gully erosion. In contrast, mean annual erosivity values reflect the risk of erosion by rain splash, overland flow and rills. By superimposing the maps of p^2/P and erosivity, a complete picture of erosion risk was obtained by Morgan (1976). With the Fourier index the contribution of rainfall in the rest of the year other than the month in which highest rainfall is there, is not taken into consideration.

Based on this index mean annual erosivity maps have been produced for the Middle East and Africa north of the equator (Arnoldus 1980) and for 16 countries of the European Union (Horvath et al. 2016).

Factorial scoring

Stocking and Elwell (1973) developed a simple scoring system for rating erosion risk in Zimbabwe. By taking a 1:1000000 base map, the country was divided on a grid system into units of 184 Km². Each unit was rated on a scale from 1 (low risk) to 5 (high risk) in respect of erosivity, erodibility, slope, ground cover and human occupation (density of population and the type of settlement). The five factor scores were summed to give a total score, which was then compared with an arbitrarily chosen classification system to categorize areas of low, moderate and high erosion risk. The scores were mapped and areas of similar risk delineated.

Factorial scoring approach being a simple approach can give general information about the vulnerability of an area to erosion risk and the areas having a high vulnerability can be assessed in more detail. Limitation in this method is its sensitivity to different scoring systems, ignoring interaction among factors, combination of factors by addition rather than by multiplication and to provide equal weightage to all factors. Binonnais et al. (2002) modified the factorial method by devising a system based on the susceptibility of the soils to crusting (four classes), the shear strength of the soil (three classes), land cover (nine classes) and rainfall erosivity (four classes). Yin et al. (2018) used factorial scoring to assess regional soil erosion risk values.

Semi-detailed assessment

Semi-detailed assessment of soil erosion hazard includes land capability classification, land systems classification and soil erosion survey.

Land capability classification

The objective of land capability classification is to divide an area of land into units with similar kinds and degrees of limitation. Various types of land capability classifications are available in different countries or geographical areas giving importance to local factors. All these land capability classifications have been developed from that given by United States Natural Research Conservation Service. The basic unit is the capability unit. This consists of a group of soil types of sufficiently similar conditions of profile form, scope and degree of erosion to make them suitable for similar crops and warranted the use of similar conservation measures.

The capability units are combined into sub-class according to the nature of the limiting factor and these, in turn, are grouped into classes based on the degree of limitation. The land capability unit is often the same as a soil series in the pedological sense, but may not be same always. In US system land is allocated into eight classes arranged from class 1, characterized by no or very slight risk of the damage to the land when used for cultivation, to class VIII, very rough land that can be safely used only for wildlife, limited recreation and watershed conservation. Class I to IV is suitable for agriculture and remaining classes are unsuitable.

The system of land capability classification was modified to cater to the needs of specific regions. A more detailed assessment of erosion risk was given by Soil Survey of England and Wales (1983) by combining the data of land capability class with rainfall erosivity and wind velocity with knowledge of the susceptibility of soils to erosion. Levin et al. (2017) provided detailed information on changes in the use of soil capability classification in the changing scenarios.

Land system classification

In land system classification landform (especially slopes), soils and the plant community is taken into consideration. This system represents the dynamic-erosion response units that reflect both the extent of erosion at any given time and its evolution over time. The land is classified into real unit termed land systems, which are made up of smaller units called land facets. Land system analysis is used to compile information on the physical environment for the purpose of resource evaluation.

Soil erosion survey

Soil erosion surveys are carried out using data from aerial photographs or satellite imageries. These surveys are helpful as they provide important information about the erosion status at a given time, change in erosion status with time and explain interrelationships between erosion features and the factors affecting them. Static surveys consist of mapping, often from aerial photographs, the sheet wash and the rills and gullies occurring in an area. Simple indices such as gully density are used to assess erosion hazard. Sequential surveys are done by comparing the results of static surveys conducted at two or more dates from the same area. In dynamic survey the factors affecting erosion for example soil type, land use, erosivity, etc., are also taken into consideration.

Soil survey maps give information on the distribution and type of erosion, erosivity, runoff, slope steepness, slope curvature, relief, soil type and land use, etc. Each component of the geomorphologic map can be digitized and stored as a separate layer in a geographic information system. Each layer can then be updated as changes occur and can be used in erosion prediction models, etc.

Satellite imageries are very useful as these provide repetitive exposures at cheaper rates as compared to aerial photographs although ground resolution of

10 to 30 m may be coarse enough to provide details of all the earth features. These can provide data on areas of bare soil to accuracies more than 25% when compared to field data (Verbyla and Richardson 1996). Satellite imagery can provide percentage vegetative cover by using Normalized Difference Vegetation Index (NDVI) (Mathieu et al. 1997). The remote sensing data will be used in future to predict the risks of erosion so that appropriate protection measures can be implemented (Panos et al. 2015).

Detailed assessment

Detailed soil erosion surveys which give information on extent and severity of erosion, are conducted at selected sites manually. These are also conducted on ground truth study for remote sensing data. Easily visible features such as the exposure of tree roots, crusting of the soil surface, formation of splash pedestals, the size of rills and gullies and the type and structure of plant cover are taken into account. Factorial scorings are used to rate the severity of erosion. For evaluating the density of rills and gullies and for tree cover large scale (usually 100 m^2), for shrub covers medium scale (usually 10 m^2) and for grass cover, crusting and depth of ground lowering, etc., small scale (usually 1 m^2) is used. Performa can be designed. As there is much seasonal variation in vegetation and soil erodibility, therefore selection of the proper time of survey and its recording is very important.

Measurement of Soil Erosion

Measurement of soil erosion is important for various objects such as monitoring soil erosion from a given area for policy makers, toinstallation of conservation measures and for research purposes, towards understanding the various processes of erosion.

Measurement of soil erosion can be done at macro-scale, meso-scale and at micro-scale both under field and laboratory conditions.

Measurement at macro-scale

Macro-scale measurement of soil erosion includes erosion by streams and rivers and the area may vary from a few hundred to a few thousand square kilometers. The purpose is to study the geographical, ecological and regional aspects of soil erosion to plan development strategies at the regional or national level. Global maps of erosion rates have been prepared by using this approach (Fournier 1960). Some of the methods used to measure soil erosion at the macro-scale are presented in brief.

Measurement from rivers

Measurement of soil erosion from river basins can be done by computing runoff from the water level, velocity, and discharge over time from a river. Velocity of channel flow can be measured using current meters, floats, dyes or traces, etc. Measurements should be done from a regular and stable stream bed area. Structures such as weirs, flumes, etc., are also constructed sometimes to control a cross sectional area. The concentration of sediments in this runoff can be measured by sampling suspended sediments and from the bed load transport. Indirect measurement of sediment load can be done using turbidity meters, neutron or gamma probes, etc. As concentration of suspended and bed load sediments vary with discharge, so calibration curves relating sediment load to discharge can be formulated. The accuracy of this method is highly dependent on the frequency of sampling.

Measurement from reservoirs

Erosion rates over a delineated watershed can be calculated if the major stream draining it passes through a well-defined reservoir. The sediments accumulated in a reservoir over the known period are converted to the erosion rate over the entire watershed. More rapid reservoir surveys can be made using an echo-sounder to obtain depth readings, an electro-distance measuring theodolite or laser theodolite to fix the position of the sounding, and a digital elevation model (DEM) to produce the contour map.

Estimation from regression models

Various empirical models have been developed to relate sediment yield to the characteristics of rainfall, runoff and watershed. The amount of sediment delivered at the watershed outlet is only a fraction of the gross erosion that occurs within the watershed. The ratio of sediment delivered at the watershed outlet to gross erosion within the watershed is called the delivery ratio. The sediment delivery ratio, which depends on number of geomorphologic and environmental factors, ranges from 5% for large watersheds of about 1000 Km^2 to about 70% for small plots of about 0.2 ha. The regression models developed to estimate soil erosion from watershed are location specific and cannot be extrapolated.

Estimation from soil surveys

Soil surveys have long been used to quantitatively estimate erosion hazard over large areas. Various indicators such as the percentage of bare grass, canopy density and density of ground cover have been used to rate soil erosion. Many regional and national maps have been developed from these databases. Aerial photography and satellite imageries are also being used for this purpose. The topic has already been discussed under erosion hazard assessment. Various reconnaissance methods of estimating soil erosion used at macro and meso-scale are discussed in next section.

Use of tracers

Radioactive isotope Caesium–137 (^{137}Cs) is the most commonly used tracer in soil erosion measurement. By measuring the isotope content of soil cores collected on a grid system the spatial pattern of isotope loading is established. This Caesium–137 fell from the atmosphere during the testing of nuclear weapons from 1950s to 1970s. Models are available to convert this information into estimates of erosion rates (Walling et al. 2002). Generally, soil erosion rates obtained using Caesium–137 compare well with measured rates from erosion plots and instrumented catchments (Theochoropoulos et al. 2003, Zhang et al. 2003, Lionel et al. 2018).

Measurement at meso-scale

The meso-scale involves the evaluation of soil erosion at the scale of farm units, i.e., from a few hectares to few hundred hectares. Various techniques used at macro-scale for example the use of radioisotopes, aerial surveys and satellite imageries can be used at the meso-scale also. Various reconnaissance methods are available which can be used for the semi-quantitative assessment of soil erosion.

Reconnaissance methods

Reconnaissance methods are ways to get a first approximation of the amount of erosion in a given situation—this approximation may be all that is needed, or it could be followed by more precise studies if required. These reconnaissance methods are cheap and simple, and need only semi-skilled staff and require little maintenance. Many measurements can be made, so these are reliable and more representative than single precise measurement. Simple techniques may also be useful as demonstrations, when the object is not to measure the amount of runoff or soil loss, but to show farmers, or extension workers, or the general public, that a lot of erosion is taking place and something should be done about it.

A common problem in all off-station field trials is the interference with the equipment by the local population. The solution is not to let it happen and then react, but to anticipate it and avoid it. This means gaining the confidence and cooperation of the local community. They should already know about the project from having been involved in its planning—if they were not, the project is starting off on the wrong foot. So, a public relations programme is required to explain what is happening, how it will help and hence secure the support of the whole local population.

The direct measurement of changes in soil level is appropriate in the case of localized erosion where rates are high and the position of the erosion can be predicted, such as steep land, which has been deforested, or cattle tracks on rangeland. It is usually not suitable for soil losses from arable land because the surface level is affected by cultivation and settlement.

Individual measurements of change in level at a single point will vary widely, but if it is an inexpensive and simple method, and a large number of points can be sampled, then a usable estimate can result. Point measurements include use of erosion pins, paint collars, pedestals, bottle tops, exposed tree root measurements and profile meters. Because there is much more spatial variation in soil erosion, so large number of point measurements give a usable estimate.

Erosion pins: This widely used method consists of driving a pin into the soil so that the top of the pin gives a datum from which changes in the soil surface level can be measured. Alternatively called pegs, spikes, stakes or rods, the pins can be of wood, iron or any other material, which will not rot, or decay and is readily and cheaply available. The pin should be a length which can be pushed or driven into the soil to give a firm stable datum: 300 mm is typical, less for a shallow soil, more for loose soil. A small diameter of about 5 mm is preferable, as thicker stakes could interfere with the surface flow and cause scour. A rectangular or square grid layout will give a random distribution of points with spacing appropriate to the area being studied.

Paint collars: An indication of large changes in level, for example in a stream bed or gully floor, can be obtained by painting a collar just above soil level round rocks, boulders, tree roots, fence posts, or anything firm and stable. Erosion reveals an unpainted band below the paint line, indicating the depth of soil removed. When painting the collar it is advisable to mask the soil with old newspaper as paint accidentally sprayed or brushed onto the soil might make it less erodible.

Bottle tops: Another simple way to record the original level is to press bottle tops into the soil surface. The depth of subsequent erosion is shown by the height of the pedestals where the soil is protected by the bottle top. This leads to the use of naturally occurring indicators of changes in soil surface level.

Pedestals: When an easily eroded soil is protected from splash erosion by a stone or tree root, isolated pedestals capped by the resistant material are left standing up from the surrounding ground. The erosion of the surrounding soil is shown to be mainly by splash rather than by surface flow if there is little or no undercutting at the base of the pedestal. Like the bottle top method, it is possible to deduce approximately what depth of soil has been eroded by measuring the height of the pedestals.

Tree roots: Exposed tree roots may offer a valid indication of change when the reason is obvious, such as erosion in a streambed below a paint collar. Very long-term rates of erosion (over several centuries) have been estimated from tree root exposure.

Profile meters: To measure small changes in surface level along a cross section such as an area with a number of parallel cattle tracks, a profile meter may be suitable. The requirement for a profile meter is to be able to set up a datum from which changes in level can be measured along a straight line and which can be re-established at the same points later to measure changes in level. Usually this takes the form of a horizontal bar with rods, which can be lowered down to the soil surface, and is the same principle as used to measure surface roughness in studies of tillage and tilth.

Volumetric measurements

Estimates of soil loss based on three-dimensional measurements of volume can be used in different ways. For erosion from rills or roads, the length of the eroding section and changes in cross-sectional area are measured. For gully erosion, usually information is needed not only on the volume lost, but also on how much the gully is increasing, so changes in length as the gully cuts back also have to be measured. The other volumetric approach is to measure or estimate the volume deposited as an outwash fan, or in a catch pit or reservoir.

Rills: Measuring the cross-section of all the rills in a sample area or along a sample transect is quick and easy, so the method is suitable for measuring change over short time periods, such as the change caused by a single heavy storm. The cross-section may be re-estimated from measurements of average width and depth if the shape is fairly uniform, or by summing the area of segments if the cross-section of the rill is irregular. The accuracy of estimates of total soil loss based only on measurements of rill erosion will depend on how much inter-rill erosion by splash and sheet wash is also occurring.

Gullies and stream banks: When the progress of gully erosion is being studied, measurements are needed both of the horizontal spread of the gully and vertical changes within the gully. To measure the surface area, and changes from cutting back or bank collapse, a rectangular grid of erosion pins is set out at an appropriate grid interval. From measurements along the grid lines from the nearest pin to the gully edge, the surface area can be plotted on squared paper. The grid lines also serve as transects for cross-sections across the gully. A string is stretched at ground level along a grid line with markers at fixed intervals of, say, 1 m. At each marker the depth is measured from the gully floor using a survey staff or a ranging rod, and the section can be plotted. The volume of soil lost from the gully is calculated and subsequent measurements will quantify the changes. Changes in a gully may be interpreted from the use of sequences of photographs. The position of the camera and the direction of the photograph must be carefully recorded. For studies of the long-term development of gullies, aerial photography can be a useful tool.

Catch pits: Simple catch pits may be used to demonstrate comparisons of soil erosion under different treatments. It is not possible to get a reliable estimate of the total soil movement unless the receiving reservoir is large enough to contain the whole flow and sediment load, but smaller pits which only catch an unknown proportion of the sediment can still be used to obtain comparative information.

A simple method for measuring relative soil movement at different points in the catchment uses 'mesh bags'. A 30 cm by 30 cm square of 5 mm mesh nylon fabric is fastened on 3 sides over the same size of 2 mm mesh. The bags are pinned to the soil surface with the open edge uphill in a line across the contour to measure horizontal variation, or up-and-down slope to measure variation down the catena. Some of the soil moved by surface flow is trapped in the mesh bag and may be dried and weighed at intervals. The method is an inexpensive and simple way of studying relative soil movement at different points in the field. As an alternative to excavating catch pits, gully check dams can be used to give an approximation of the effect of different treatments in their catchments.

Direct measurements

Direct measurements are perhaps the most accurate way of measuring soil erosion, but also the most laborious and time consuming. They involve collecting deposited materials and taking volumetric and weight measurements. It includes establishing bounded runoff plots to collect surface runoff, with a flow-collecting device at outlet. Detailed field studies to measure soil erosion are usually conducted for research purposes to study basic soil erosion processes or to evaluate the effect of soil conservation practices on soil erosion. These studies are either done under natural rainfall conditions or under artificial rainfall conditions using a rainfall simulator in the field or laboratory conditions.

Field measurement with tracers

Measuring the dilution of mobile substances, which can be entrained by water if added to soil over time, enables volumetric rates of surface erosion or deposition to be calculated. An enormous variety of substances have been tried which includes soluble dyes and radionuclides with greater success than others.

Measurement from stream discharge

It is an indirect method of measuring soil erosion by collecting suspended sediment and/or bed load with a sampler. By recording the stream flow, sediment concentrations may be converted to volumes eroded from upstream catchment.

Measurement at micro-scale

The area of study may vary from a fraction of a meter to a few hundred square meters. The study may be done under natural *in situ* conditions or under laboratory conditions. Natural or simulated rainfall can be used for this purpose.

Measurement from field plots

To study the factors affecting erosion, bounded plots are employed at permanent research or experimental stations. These plots are installed with specific dimensions and both runoff and soil losses are monitored. Standard USLE plots (22 m long and 1.8 m wide with 9% slope) are most widely used. These plots are large enough to represent the combined process of rill and inter-rill erosion. From small plots a sediment sub-sample is taken manually from the collected runoff or part of the runoff and then the sediment yield is calculated based on total runoff volume. An automatic sediment sampler can be installed to extract samples of the runoff at regular intervals during the storm and the time at which each sample is taken can be recorded. Although the bounded runoff plot gives probably the most reliable data on soil loss per unit area, there are several sources of error involved with its use. There may be overflow during extreme events, the tanks floating out of saturated ground, runoff entering the top of the plots, the tops in the collecting tanks being left open, damage from termites, silting of the collecting trough and pipes leading to tanks, etc.

Measurements from small plots

The size of runoff plots used depends on the objectives of study. The plot size is often less than 5 m^2 if the objective is to compare simple treatments, e.g., effect of residue mulch on water runoff and soil loss. Plots of about 1–2 m^2 size will allow investigations into infiltration and effects of rain splash. For relative erodibility of soil and for comparison of ground covers, etc., plots must be at least 10 m long for studies of rill erosion. Small plot experiments should have objectives complementary to field plot experimentation or application. Small plots are also used to develop or verify basic operating equations that govern the physical processes of soil erosion for modeling purposes.

Laboratory Measurements of Soil Loss

The relative magnitude and trends in same processes at micro-scale can be evaluated under controlled conditions in laboratory. In laboratory studies there are problems of scale in the scale of the experiment, influence of boundary effects and the extent to which field conditions are simulated.

Splash by raindrop effect and determination of soil erodibility have been extensively studied under laboratory conditions. Edge effects are greatest with small plots. For splash measurement under laboratory conditions, a border area is necessary to ensure that as much splash that will enter the plot, leaves the plot. Small laboratory plots must have a bottom to hold the soil. An open bottom made of screens overlain by cloth often is used to allow free passage of soil water although this condition does not resemble true field conditions. Under laboratory conditions disturbed soil samples are used. Soil only from the plow layer in the field is taken, air-dried and sieved through a large screen and packed in the plot container in layers. This disturbance of soil may not give true picture of field conditions.

Use of Gerlach troughs

Gerlach (1966) used simple metal gutters, 0.5 m long and 0.1 m broad, closed at the sides and fitted with a movable lid. A collecting bottle is attached with outlet pipe from the base of the gutter. Two or three gutters are placed side by side across the slope and groups of gutters are installed at different slope lengths in such a way that there is a clear run to each gutter from the slope crest. Soil loss is calculated from the sediments collected in each gutter. The sediments are assumed to be uniformly contributed by the land area and accuracy of measurements depends on the number of gutters used.

Soil erosion from micro watersheds can be measured to study the effect of various treatments like contouring, strip cropping and gully treatment, etc., because these studies are not feasible in small plots. In contrast to the plots used to study isolated parts of the erosion process, small watersheds allow for the study of the entire process.

Splash erosion measurements

Splash erosion has been measured in the field by means of splashboards or small funnels or bottles. These are inserted in the soil to protrude 1–2 mm above the ground surface, thereby eliminating the entry of overland flow, and the material splashed into them is collected and weighed. An field splash cup is also used where a block of soil is isolated by enclosing it in a central cylinder and the material splashed out is collected in a surrounding catching tray. Because the quantity of splashed material measured per unit area depends upon the diameter of the funnels and cups, the following corrections has to be applied to determine the real mass of particles detached by splash:

$$MSR = MSe^{0.054D}$$

where MSR is the real mass of splashed material per unit area (g cm^{-2}), MS is the measured splash per unit area (g cm^{-2}), and D is the diameter of the cup or funnel (cm). Scholten (2011) introduced a new splash cup based on Ellison's archetype that reliably and accurately measures kinetic energy as a function of sand loss under a large variety of conditions. The developed cup, known as, Tübingen splash cup (T splash cup) is relatively easy to operate under harsh field conditions, and can be used in experimental designs with a large number of plots and replications at reasonably low costs. The splash cups have been calibrated in combination with a laser distrometer using a linear regression function with $r^2 = 0.98$.

Rainfall Simulation

Rainfall simulation, a device to produce rainstorms of desired characteristics, has been widely used as a research tool in soil erosion studies because of the unpredictable, infrequent and random nature of rainfall. The major advantages of rainfall simulator research are four fold: it is more rapid, more efficient, more controlled and more adaptable than natural rainfall research. The disadvantages of rainfall simulators are cost and time required to construct a suitable rainfall simulator and the difficulty of simulating natural rainfall characteristics. Important design requirements of simulators include rainfall intensity, raindrop size, drop size distribution, drop velocity at impact, and kinetic energy of rainfall. Rainfall simulators can broadly be classified into two groups, i.e., those involving nozzles from which water is forced at a significant velocity by pressure and those where drops form and fall from a tip starting at zero velocity. Detailed information on use of rainfall simulators in soil erosion is available in literature (Shrivastva and Das 1998). Rainfall intensity, length of simulated rainstorms and sequence of rainstorm can be varied as per requirement of the study. Mhaske et al. (2019) designed rainfall simulator for soil erosion studies in laboratory.

Rainfall simulation under field conditions: In recent years considerable use has been made of rainfall simulators in the field conditions. Natural runoff plots have been virtually replaced by these simulated studies as a research tool. Field rainfall simulators provide the advantages of field conditions for soils, slopes and plant cover, all of which are difficult to reproduce in the laboratory, with the benefits of a repeatable storm. Several designs for simple, portable simulators have been produced (Cerda et al. 1997) with the ability to generate rainfall at intensities between 40 and 120 mm h^{-1}.

Runoff simulation: In small soil plots, the rainfall simulator may be supplemented by a device to supply a known quantity of runoff at the top of the plot. Sediment can also be added to the runoff upslope of the test soil. This type of simulation is usually used to study the rill erosion under controlled conditions.

Rill Erosion Measurement

Series of transects, 20–100 m long, across the slope and positioned one above the other is used to assess soil loss from rill erosion. The cross-sectional area of rills is determined along two successive transects. The average of the two areas multiplied by the distance between the transects, gives the volume of the material removed. Weight of soil loss is calculated from the volume of soil and bulk density. This soil loss is from the area between two transects. There is difficulty in measuring temporal changes in cross-sectional areas of rills as the reference level is also changed so it underestimates rill erosion by 10–30%. Chen et al. (2016) used a volume replacement method to estimate rill erosion. Zhang et al. (2019) used a soil erosion prediction model for rill erosion measurements.

Gully Erosion Measurement

The technique used to study rill erosion can be used for small gullies also. For large gullies, sequential surveys using aerial photography are more suitable. Rates of gully erosion are calculated from the differences in elevation between DEM's at different dates (Betts et al. 2003, Baily et al. 2003). More recent surveys of gullies use large scale (1:10000) aerial photography to construct high-resolution digital elevation models (DEMs) (Nicholas et al. 2017).

References

Arnoldus, H.M.J. 1980. An approximation of the rainfall factor in the universal soil loss equation. pp. 127–132. *In*: De Boodt, M. and D. Gabriels (eds.). Assessment of Erosion. Wiley Chichester.

Baily, B., P. Collier, P. Farres, R. Inkpen and A. Pearson. 2003. Comparative assessment of analytical and digital photogrammetric methods in the construction of DEM's of geomorphological forms. Earth Surf. Process. Landf. 28: 307–320.

Betts, H.D., N.A. Trustrum and R.C. Rose. 2003. Geomorphic changes in a complex gully system measured from sequential digital elevation models, and implications for management. Earth Surf. Process. Landf. 28: 1043–1058.

Binonnais, Y., C. Montier, M. Jamagne, J. Daroussin and D. King. 2002. Mapping erosion risk for cultivated soils in France. Catena 46: 207–220.

Cerda, A., S. Ibanez and A. Calvo. 1997. Design and operation of a small and portable rainfall simulator for rugged terrain. Soil Tech. 11: 163–170.

Chen, X.Y., Y. Zhao, H.X. Mi and B. Mo. 2016. Estimating rill erosion processes from eroded morphology in flume experiments by volume replacement method. Catena 136: 135–140.

Fournier, F. 1960. Climat et erosion: la relation entre l'erosion du sol par l' eau et les precipitatousatmospheriques. Presses Universitaires de France, Paris.

Gerlach, T. 1966. Wspolczesbyrozwaystokow w dorzeczugorzecz'sgornezoGrajcarka (BeskidWysok-KarpatyZachodnie). PraceGeograf. IG PAN 52.

Horvath, C., R. Kinga and O. Rosian. 2016. Assessing rainfall erosivity from monthly precipitation data. AerulsiApa. Componente ale Mediului; Cluj-Napoca: 109–116.

Levin, M.J., R. Dobos, S. Peaslee, D.W. Smith and C. Seybold. 2017. Soil capability for the USA now and into the future. *In*: Field, D.J., C.L.S. Morgan and A.B. McBratney (eds.). Global Soil Security. Progress in Soil Science. Springer, Cham.

Lionel, M., C. Bernard, A. Lee, Y. Zhi, E. Fulajtar, G. Dercon, M. Zaman, A. Toloza and L. Heng. 2018. Promoting the use of isotopic techniques to combat soil erosion: An overview of the key role played by the SWMCN Sub programme of the Joint FAO/IAEA Division over the last 20 years. 29: 3077–3091.

Mathieu, R., C. King and Y. LeBissonnais. 1997. Contribution of multi-temporal SPOT data to the mapping of a soil erosion index. The case of the loamy plateaux of Northern France. Soil Tech. 10: 99–110.

Mhaske, S.N., K. Pathak and A. Basak. 2019. A comprehensive design of rainfall simulator for the assessment of soil erosion in the laboratory. Catena. 172: 408–420.

Morgan, R.P.C. 1976. The role of climate in the denudation system: a case study from peninsular Malaysia. Malay. Nat. J. 28: 94–106.

Nicholas, R., G. John, D. Armston, J. Muir and I. Stiller. 2017. Monitoring gully change: A comparison of airborne and terrestrial laser scanning using a case study from Aratula, Queensland. Geomorphology 282: 195–208.

Panagos, P., K. Meusburger, C. Ballabio, P. Borrelli and C. Alewell. 2014. Soil erodibility in Europe: A high-resolution dataset based on LUCAS. Science of Total Environment 479-480: 189–200.

Panos, P., P. Borrelli, K. Meusburger, C. Alewell, E. Lugato and L. Montanarella. 2015. Estimating the soil erosion cover-management factor at the European scale. Land Use Policy 48: 38–50.

Scholten, T., C. Geibler, J. Goc, P. Kuhn and C. Wiegand. 2011. A new splash cup to measure the kinetic energy of rainfall. J. Plant Nutr. Soil Sci. 174: 596–601.

Shrivastava, R.K. and G. Das. 1998. A review of rainfall simulators for soil erosion research studies. Indian J. Soil Conserv. 26: 26–80.

Soil Survey of England and Wales. 1983. Soil map of England and Wales. 1: 250000 map sheets.

Stocking, M.A. and H.A. Elwell. 1973. Soil erosion hazard in Rhodesia. Rhodesian Agric. J. 70: 93–101.

Theochoropoulos, S.R., H. Floron, D.E. Walling, H. Kalantzakos, M. Christon, P. Tountos and T. Nikolaou. 2003. Soil erosion and deposition rates in a cultivated catchment area in central Greece, estimated using the ^{137}Cs technique. Soil Till. Res. 69: 153–162.

Verbyla, D.L. and C.A. Richardson. 1996. Remote sensing clear cut areas within a forest watershed: comparing SPOT, HRV Panchromatic, SPOT HRV multispectral, and Landsat Thematic Mapper data. J. Soil Water Conserv. 51: 423–427.

Walling, D.E., Q. He and P.G. Appleby. 2002. Conversion models for use in soil erosion, soil redistribution and sedimentation investigations. pp. 111–164. *In*: F. Zapata (ed.). Handbook for the Assessment of Soil Erosion and Sedimentation Using Environmental Radio Nuclides. Kluwer, Dordrecht.

Wischmeier, W.H. and D.D. Smith. 1978. Predicting rainfall erosion losses. USDA Agricultural Research Service Handbook 537.

Yin, S., Z. Zhengyuan, W. Li, L. Baoyuan, X. Yun, W. Guannan and L. Yishan. 2018. Regional soil erosion assessment based on a sample survey and geostatistics. Statistics Publications: 133.

Zhang, P., W. Yao, G. Liu and P. Xiao. 2019. Experimental study on soil erosion prediction model of loess slope based on rill morphology. Catena. 173: 424–432.

Zhang, X., Y. Zhang, A. Wen and M. Feng. 2003. Assessment of soil losses on cultivated land by using the ^{137}Cs technique in the upper Yangtze River basin of China. Soil and Till. Res. 69: 99–106.

Chapter 5

Measures to Control Soil Erosion

Abrar Yousuf,[1], Jonas Lenz[2] and Eajaz Ahmad Dar[3]*

Introduction

Soil is a critical resource for the future of mankind. It has to be protected and enhanced. Instead, more than half (52%) of all fertile, food-producing soils globally are now classified as degraded, many of them severely degraded (UNCCD 2015). Soil degradation is the decline in any or all of the characteristics which make soil suitable for producing food. Soil degradation occurs through the deterioration of the physical, chemical and biological properties of soil that results in soil compaction, salinization, acidification, and soil loss from wind and water erosion. Soil degradation is a severe environmental problem, affecting about 1100 million ha worldwide (56% of the total area affected by human-induced soil degradation). Almost 80% of the terrain affected by water erosion has a light to moderate degree of degradation. Among the major continents, Africa ranks second in the severity of soil erosion after Asia (Oldeman 1992). The latest reference in this regard is given by United Nations which states that the majority of the world's soil resources are

[1] Regional Research Station (Punjab Agricultural University), Ballowal Saunkhri, SBS Nagar, 144521, India.
[2] TU, Bergakademie Freiberg, Soil and Water Conservation Unit, Freiberg, Germany.
  Email: jonaspunktlenz@gmail.com
[3] Subject Matter Specialist (Agronomy), KVK- Kargil (SKUAST-K), J&K, 194103, India.
  Email: dareajaz9@gmail.com
* Corresponding author: er.aywani@gmail.com

in only fair, poor or very poor condition (FAO and ITPS 2015). Deforestation, tillage, inappropriate cultivation practices and over grazing are among the major causes of soil erosion. The FAO led Global Soil Partnership has reported that 75 billion tonnes (Pg) of soil are eroded annually from arable lands worldwide, which equates to an estimated financial loss of US$400 billion per year (GSP 2017). A recent study by the Economics of Land Degradation Initiative (ELD) calculated that global soil degradation costs us between US$6.3 and US$10.6 trillion (£4.4 to £7 trillion) per year. The ELD study also estimated that US$480 billion (£317 billion) could be generated by enhancing carbon stocks in soils, and that by adopting more sustainable farming practices increased crop production worth an US$1.4 trillion (£900 million) could be achieved (ELD 2015).

Soil erosion is common in all areas of the world, but developing countries suffer more because of the inability of their farming populations to replace lost soils and nutrients (Mohamed 2015). Soil erosion impacts food security in developing countries and these countries are further confounded by harsh climate (e.g., frequent drought or flooding) and poor socio economic and political stability (Blanco-Canqui and Lal 2010). Therefore, it is important to conserve the soil to sustain life on earth and to ensure the food security in the world. There are three main principles to control the soil erosion: use land according to its capability, protect the soil surface with some form of cover and control runoff before it develops into an erosive force. There are two different ways to control soil erosion depending on the topography of the land: (1) Agronomic Measures and (2) Mechanical/Engineering Measures. Agronomic measures are considered the first line of defense against the soil erosion. These measures are more economical, effective and long lasting. On the other hand, mechanical or engineering measures are used to control the soil erosion immediately. They are considered a second line of defense. Generally, engineering measures are employed only when the agronomic measures are not sufficient to control the soil erosion. Engineering measures are generally expensive as they involve the construction of different types of structures to control the soil erosion. The appropriateness of a particular adaptation strategy is highly dependent on time and place as they are influenced by the cultural and indigenous observations and practices (Obert et al. 2016).

Soil Conservation Strategies

Cultivated lands

The ground cover on agricultural lands can be of different types, e.g., varying in height, density, and canopy structure, organic or inorganic. When a forest or other land use is converted to agricultural land use, the forest or grass cover is removed. This change in land use is generally associated with increased soil erosion. When this land use change is on steep slopes, soil erosion may be very high. When the practices like cultivation up and down the hill, inadequate use of fertilizers and manures, use of heavy machinery are used, it results in a steep rise in soil erosion. Management of soil erosion, which includes application of agronomic and soil management principles, can be efficiently done by supportive mechanical methods of erosion control.

Conservation strategies are aimed at establishing and maintaining good ground cover. A complete cover within 50 cm of the soil surface is extremely effective in minimizing raindrop impact. The effectiveness of the cover however, decreases with increasing height above the ground surface. The tree canopy 10 to 20 m above the ground surface loses its effectiveness because the coalescing drops falling from the canopy are generally big, attain terminal velocity and have high impact energy.

Crops grown in rows, tall tree crops and low growing crops with large leaves afford least protection of the soil. A good crop cover is essential at the time of year when high intensity rainstorms are expected. Quick growing crops may be viewed as soil-conserving crops. Unfortunately, the crops, which prevent erosion problems, are normally high value crops or are food crops. The challenge is to develop soil conservation strategies that will allow all these crops to be grown productively in a short period to meet the immediate need of the farmers and sustainability in the long term, and also so as not to deplete the soil resources for future generations.

Soil conservation measures must be both technically sound and socially and economically acceptable to the farmers. It is recognized that strategies for soil conservation must rely on improving traditional systems, instead of imposing entirely new techniques from outside (Roose 1992).

Pasture lands

These comprise areas of improved pasture where grasses and legumes suited to the local soil and climatic conditions are planted and managed by regular applications of fertilizers and organic manures, as well as areas of rangeland composed of native grasses and shrubs. Since grass provides a dense cover, close to the soil surface, it is a good protector of the land against erosion. Erosion problems arise only when this cover is removed through overgrazing although they can be exacerbated by drought and excessive burning. Agronomic measures are usually adopted to control erosion in pasture lands. Controlled grazing and growing of erosion resistant grasses and shrubs can be done to combat erosion. Erosion resistant grasses are characterized by vigorous growth, tolerance to drought and poor soils, palatability to livestock and resistance to the physical effect of trampling. Traditional growing systems are often well adapted to the local conditions of climate, soils and vegetation, making use of rotational grazing on a nomadic basin. All traditional grazing systems are under pressure because of an increase in human and livestock population. Usually pasture lands belong to the community, whereas the livestock belongs to individuals hence leading to conflicts.

Forest lands

Forest lands usually have a multi-tier canopy which protects the land from erosion. The addition of sufficient quantities of organic matter leads to an improvement in the physical condition of soil such as infiltration, aggregation, water holding capacity, etc. Low runoff rates and the protective role of the litter layers on the surface of the soil produce low erosion rates in forest lands. Erosion is abruptly increased when forests are cleared for agriculture. In forest conditions destruction of trees and shrubs by grazing, cutting of trees for firewood and logging operations cause erosion, and the growing of quick-growing tree species for firewood can be a strategy to reduce the cropping of forest trees.

Logging operations if done using mechanical methods cause more erosion than manual clearings. Forest removal causes the loss of shear strength gradually; following the decay of root systems, which induces a risk of landslides.

Rough lands

Rough lands are usually located in hilly and mountainous terrain with shallow stony soils and steep slopes or in sand dunes. These are too marginal that they cannot be used for agriculture or forest land use and usually used for recreational purposes. In these lands, the overuse of paths and tracks results in a reduction in overall vegetative cover, compaction of soil and changes in soil moisture. Erosion control strategies in these areas include exclusion of people, use erosion-resistant plant species, improving drainage and soil strength. The plant species selected for re-vegetation of rough lands should ideally be local.

Agronomic Measures to Control Soil Erosion

The agronomical measures include growing vegetation on the land with fewer slopes to cover them and to control the erosion from such lands. Plant cover protects the soil against erosion by reducing water runoff (Rey 2003, Puigdefabregas 2005, Durán et al. 2006) and by increasing water infiltration into the soil matrix (Wainwright et al. 2002). Plants shelter and fix the soil with their roots (de Baets et al. 2007) and reduce the energy of raindrops with their canopy (Bochet et al. 1998). Also, vegetation can act as a physical barrier, altering the sediment flow at the soil surface (Martínez et al. 2006). The principle behind the agronomic measures is that the vegetation covers the soil surface and protects it from the erosive nature of rainfall and hence reduces the splash erosion. The vegetation on the soil increases the porosity and helps in increasing the infiltration rate of the soil and thereby reducing the surface runoff. This results in a decrease in the rill and inter-rill erosion. It also reduces the runoff velocity which results in lesser detachment of the soil particles and hence lesser soil erosion. The agronomic measures are also helpful in reducing the wind erosion as they act as a buffer against the abrasive nature of the winds. Agronomic measures include contour cultivation, strip cropping, tillage practices and soil management practices. Edward and Simon (2003) noted that conservation, minimum tillage, mulches and cover crops prevent runoff initiation by intercepting raindrops in a handbook of processes and modeling in the soil-plant system. Dimelu et al. (2013) studied the soil erosion conservation practices in Enugu, and his results showed that the soil conservation techniques used as adaptive measures were crop rotation, mulching, liming, contour bonds and terracing.

Contour cultivation

In contour cultivation, all farming operations such as ploughing, sowing, tillage, etc., are done along the contour or against the natural slope of the field. This is a very simple technique to conserve the soil and water in the field. Contour cultivation not only conserves the rain water within the field only but it also retards the flow of the runoff water. This results in less soil erosion. Contour ploughing builds a barrier against rainwater runoff which is collected in the furrows and results in higher infiltration. Contour ploughing is especially important at the beginning of the rainy season when biological conservation effects are poor. This method is effective on moderate slopes. Tillage and planting operations follow the contour line to promote positive row drainage and reduce ponding. Also, by increasing the soil surface roughness, contour ridging results in rainwater ponding in the furrow area, which reduces runoff velocity, increases infiltration, and reduces soil erosion (Liu et al. 2014). In addition, nutrients (e.g., nitrogen and phosphorus) in runoff are retained better in contour ridge tillage compared with up and downslope tillage (Ma et al. 2010, Liu et al. 2014). In dry areas, contour farming increases crop yield by increasing infiltration and retaining water. The effectiveness of contour ploughing decreases with an increase in slope gradient and length, rainfall intensity and erodibility of the soil. The effectiveness of contour farming in controlling soil erosion varies with the soil texture, land slope and crop cover.

Strip cropping

Strip cropping is the system of growing alternate strips of erosion permitting crops (row crops such as maize, jowar, bajra, cotton, etc.), and erosion resisting crops (close growing crops such as green gram, black gram, moth, groundnut, etc.), in the same field. This practice reduces the velocity of runoff and checks the eroded soil from being washed away. Strip cropping is essential for controlling the run-off erosion and thereby maintaining the fertility of the soil and is now universally recognized. The effectiveness of the strip cropping in controlling runoff and soil erosion is due to following reasons: it reduces the surface runoff and increases the infiltration in to the soil. The reduction in surface runoff is due to the obstruction to the flow caused by the crops and thus allowing the runoff more time to infiltrate into the soil. In addition to controlling the runoff and soil erosion, it is also effective in enhancing the soil fertility.

Strip cropping is practiced in four different ways: (1) Field strip cropping, (2) Contour strip cropping, (3) Buffer strip cropping, and (4) Wind strip cropping.

Field strip cropping

Field strip cropping includes growing the different crops in uniform strip widths across the normal field slope irrespective of the field's contours. This method of strip cropping is practiced on regular slopes and with the high rates of infiltration.

Contour strip cropping

Contour strip cropping includes growing different crop strips along the contours at the direction perpendicular to the normal slope of land. The alternate strips of the crops include erosion permitting and erosion resisting crops. The width of the strips depends on number of factors such as the topography of the area, soil characteristics, rainfall characteristics, etc. The crops may be rotated after some time in order to maintain the proper soil health. This type of strip cropping is practiced across the slope in order to flow and retard the flow of the runoff over the land surface.

Buffer strip cropping

The buffer strip cropping system is practiced on lands which are facing severe erosion and do not fit into a regular rotation. In this system, permanent strip of grasses or legumes or a mixture of both is laid in strips to protect the soil from erosion. The width of the strips may or may not be uniform.

Wind strip cropping

Wind strip cropping includes laying the crop strips at a direction perpendicular to the direction of the prevailing winds irrespective of the direction of land slope. The purpose of this cropping system is to control the wind erosion rather than the water erosion.

Wind strip cropping is the production of the regular farm crops in long, relatively narrow, straight, parallel strips placed crosswise of the prevailing winds without regard to the contour of the land. Wind strip cropping is more effective in retarding wind erosion than contour strip cropping, but usually has little value in conserving water (Kell and Brown 1937).

Strip cropping combined with contour cultivation has been proved to be very economical and effective and one of the most practical and simplest means of conserving soil and water on cultivated land. Contour strip cropping divides the length of the slope and checks the velocity of runoff water, filters out the soil being carried off, and increases the infiltration of rain water by the soil. Strip cropping provides for a larger number of small fields and encourages the use of a proper crop rotation system and helps maintain a balance of soil-building and harvested crops. It can be practiced at practically no expense and the cost of maintenance is very low.

Mulching

Mulching is the practice of covering the soil surface with the plant residues or other suitable material for example plastics. Mulching modifies the micro-climate around the crop by affecting the soil moisture and soil temperature. It also reduces the weed growth and enhances nutrient availability. In addition, mulching has been found to be very effective in controlling soil erosion as it affects the erosion process at different stages. Firstly, it dissipates the kinetic energy of the falling drops and prevents the direct contact of the erosive raindrops on the soil surface thereby decreasing splash erosion. Secondly, it obstructs the flow of the runoff water preventing it from attaining the erosive velocity thereby preventing the sheet or inter-rill erosion. It also improves the infiltration capacity of the soil. Crop residues, straw materials, grasses, sawdusts, compost, gravel, crushed stone, plastics etc. are used as mulching materials.

Vashisht et al. 2013 evaluated effect of three different mulches (sugarcane trash, twigs and leaves of basooti and subabul) on soil erosion and the grain yield of maize, and the carryover residual soil moisture for the sowing of rain fed wheat. It was observed that lower runoff and soil loss was recorded in the mulched plots than the un-mulched (control) plots. Also, more water was conserved in the profiles with mulched treatments as compared to the control; during the crop growth and at the harvest of maize.

Types of mulching materials

Mulching material can be classified in two broad groups:

Organic Mulches: Organic mulches are derived from plants. They include crop residues, straw materials, grasses, saw dusts, compost, leaves, twigs, etc. They decompose relatively quickly, sometimes in one season, enriching the soil with organic matter and nutrients.

Inorganic Mulches: Inorganic mulches include stone mulch, soil mulch, gravel and pebbles. Stone mulching is a practice of spreading stones on the soil surface to conserve the soil moisture. It also reduces wind erosion. This type of mulching is mostly practiced in arid regions. Soil mulch involves establishing a thick layer of loose and dry soil on the soil surface. It reduces the capillary loss of water from the lower layers as it breaks the contact with the moist soil layer, increases the non-capillary pore spaces and resistance to wetting.

Tillage practices

Tillage is the mechanical manipulation of soil with tools and implements for obtaining conditions ideal for seed germination, seedling establishment and growth of crops. The purpose of tillage practices is to prepare a good seed bed which helps the germination of seeds and creates conditions in the soil suited for the better growth of crops. It is an important operation which controls the weed growth, maintains the infiltration capacity and soil aeration.

Types of tillage

Conventional tillage

Conventional tillage involves primary tillage to break open and turn the soil followed by the secondary tillage to obtain seed beds for sowing or planting. In conventional tillage, continuous use of heavy ploughs creates a hard pan in the subsoil, results in poor infiltration. It makes the soil more susceptible to run-off and erosion. It is capital intensive and increases soil degradation. To avoid these ill effects, modern concepts on tillage are being followed.

Conservation tillage

Conservation tillage is the practice of ploughing the fields with a lesser number of passes over the entire land or ploughing only in the required space of the land. It is a system of tillage in which organic residues are not inverted into the soil such that they remain on surface as protective cover against erosion and the evaporation losses of soil moisture. The residue left on the soil surface interferes with seed bed preparation and sowing operations. The advantages of conservation tillage are: (a) Energy conservation through reduced tillage operations (b) Improve the soil's physical properties (c) Reduce the runoff water from fields.

Different types of conservation tillage are minimum tillage, no tillage and mulch tillage

Minimum tillage: Minimum tillage reduces the tillage operations to a minimum for seed bed preparation. In minimum tillage, tillage and the sowing operation are combined in one operation. This type of tillage creates a coarse soil surface and fine lumps of soil between rows and hence increasing the infiltration capacity of the soil. This method reduces the runoff and soil erosion from the fields.

No-tillage/Zero tillage: No tillage involves growing the crop in the residues of the preceding crop without any soil manipulation or the seed bed preparation. The soil surface in no-tillage is left undisturbed. This type of tillage is applicable for coarse textured soils having good internal drainage, favorable initial soil structure and an adequate quantity of crop residue as mulch. The no-tillage system considerably reduces the soil erosion.

Mulch tillage/Stubble tillage: Mulch tillage is the practice wherein the soil is protected from erosion either by growing the new crop or by leaving the crop residue on the fields during the fallow periods. The soil is made cloddy with the help of crop stubbles/residues. It has been reported that the mulch tillage is effective in controlling the soil erosion and conserving the soil moisture.

Vegetative barriers

Vegetative barriers are alternative biological measures, which have been shown to effectively conserve soil and water by moderating the surface runoff and allowing the increased infiltration time. Vegetative barriers are narrow strips (1–3 feet wide) of stiff, erect densely growing plants, usually grasses, planted across the slope perpendicular to the dominant slope. Vegetative barriers retard and reduce the surface runoff, control soil erosion and trap sediments at the bottom of the fields. Vegetative barriers inhibit the flow of water because of their dense concentration of thick stems, thus slowing and ponding water and causing sediments to deposit back in them (Ramajayam et al. 2007). Over time these deposits can develop into benched terraces. These barriers function to diffuse and spread the water runoff so that it slowly flows through them without erosion. Dass et al. (2011) studied the effect of different vegetative barriers planted in combination with a trench-cum-bund, on runoff, soil loss, nutrient loss, soil fertility, moisture retention and crop yield in the rain fed uplands in the Kokriguda watershed in southern Orissa. Singh et al. (2017) evaluated the effect of five vegetative barriers namely Vetivar grass (*Vetiveria zizanoides*), Bhabbar grass (*Eulaliopsus binata*), Kanna grass (*Sachharum munja*), Subabul (*Leucaena leucocephala*) and Napier bajra hybrid (*Pennisetum purpureum X typhoides*) on runoff, soil loss and crop parameters at the research farm of Regional Research Station, Ballowal Saunkhri, Punjab. India.

Mechanical Measures to Control Soil Erosion

Mechanical/engineering measures are an important component of soil and water conservation. They are used to control soil erosion where the agronomic measures are not successful in controlling the soil erosion. They are mostly used on high sloped lands where the velocity of the runoff water is very high. Mechanical measures for erosion control require a proper design based on the rainfall characteristics, soil characteristics, topography of the area and land use. While planning, mechanical measures, the following approaches are followed:

a. To increase the infiltration opportunity time, so as to increase the infiltration of runoff into the soil.
b. To reduce the slope length into series of small drops, so as to prevent the runoff from attaining the erosive or threshold velocity.
c. To protect the soil from getting eroded due to the runoff water.

Bunding

Bunds are embankment like structures constructed across the slope. Bunds break the long slope length into smaller ones preventing the runoff from attaining the erosive or threshold velocity. Bunds obstruct the flow of runoff and thus control the soil erosion. Generally, no crops are grown on the bunds except for some grasses. Bunding is usually practiced for the lands where the slopes vary from 2–10 percent. Bunding helps in increasing the soil moisture thereby increasing the crop yields. Raes et al. (2006) found that bunding results in a higher yield of low land rice in Tanzania. Gebreegziabher et al. (2009) provides evidence of a positive effect of contour bunds on water utilization and soil conservation. Kato et al. 2011 also reported that contour bunds have the potential to increase the crop yields in highlands. According to a study conducted by Adimassu et al. (2012), soil bunds brought about a significant reduction in runoff and soil loss. Plots with soil bunds reduced the average annual runoff by 28 per cent and the average annual soil loss by 47 per cent. Consequently, soil bunds reduced the losses of soil nutrients and organic carbon.

Bunds are known as contour bunds when constructed on the contour, and graded bunds when some grade (slope) is given to them. The type of bund depends on rainfall, soil type, topography and purpose of making the bund. Contour bunds are constructed in low rainfall areas where the annual rainfall is less than 600 mm and soils are relatively permeable. Graded bunds are constructed in high rainfall areas (more than 600 mm) and soils are less permeable. Graded bunds are designed to dispose of the excess runoff safely without causing the soil erosion. Graded bunds can have a uniform or variable grade. Bunding is not practiced in clay or deep black soils because these soils develop cracks in hot weather season. The different types of bunds are defined below:

Side bunds

Side bunds are constructed at the extreme ends of the contour bund. They are constructed along the slope.

Supplemental bunds

Supplemental bunds are constructed between the two contour bunds in order to minimize the horizontal spacing between the contour bunds. They are constructed against the slope.

Lateral bunds

Lateral bunds are constructed along the slope in between the two side bunds in order to protect the bunds against breaching due to the accumulation of runoff.

Shoulder bunds

The shoulder bund is constructed at the outer edge of the outward sloping terraces in order to retain the runoff water within the terrace. It also provides stability to the terrace.

Marginal bunds

Marginal bunds are constructed along the margin of the property/entity such as a watershed, field, river, etc., to demarcate their boundary.

Design specifications for bunding

1. Spacing of bund

The spacing between the two consecutive bunds should be such that it breaks the slope length to such an extent that runoff water does not attain the erosive velocity. It primarily depends on rainfall characteristics, slope and soil type. The spacing between the bunds is expressed as vertical interval.

Ramser developed a relation to calculate the vertical interval between the two bunds for sub humid areas having highly permeable soils. According to this relation,

$$VI = 0.3\left(\frac{S}{3} + 2\right)$$

where,

VI is vertical interval (m) and S is slope (percent)

For high rainfall regions, $VI = \frac{30s}{3} + 60$

For low rainfall regions, $VI = \frac{30s}{2} + 60$

where VI is vertical interval (cm) and S is slope (percent)

Cox developed a relation to calculate the vertical interval by taking into account the effect of infiltration and rainfall amount. According to Cox's relation,

$$VI = 0.3 \ (XS + Y)$$

where VI is vertical interval (m), X is rainfall factor and Y is the infiltration rate and crop cover factor, S is land slope (percent). The values of X and Y are given in Table 1 and 2.

The horizontal spacing of bunds is based on the slope of the land. The following can be used to calculate the horizontal spacing:

$$HI = \frac{VI}{S} \times 100$$

where HI is horizontal interval (m), S is land slope (percent)

Table 1. Values of rainfall factor (X).

Rainfall	Annual rainfall (mm)	X
Scanty	640	0.8
Moderate	640–900	0.6
Heavy	> 900	1.0

Source: Suresh 2016

Table 2. Values of Y based on infiltration rate and crop cover.

Infiltration rate	Crop cover during erosive period of rain	Y
Below average	Low cover	1.0
Average or above	Good cover	2.0
One of the above factors favorable and other unfavorable		1.5

Source: Suresh 2016

2. Depth of ponding water

The depth of water ponding behind the bund is calculated using the formula:

$$h = \sqrt{\frac{Re \times VI}{50}}$$

where h is depth of ponding water behind the bund

Re is rainfall excess (cm) for 10 years recurrence interval

VI is vertical interval (m)

The total height of the bund (H) is given as:

H = Depth of ponding water (h) + Depth of water over outlet + Freeboard as 25% of h

3. Height of bund

The height of the bund is calculated on the basis of the amount of water to be intercepted by the bund. The height of bund should be such that the runoff does not flow over the bund. The height of bund is given as:

$$H = \sqrt{\frac{3 \times HI \times S}{50}}$$

where H is height of bund, HI is horizontal interval between the bunds and S is slope (percent).

A freeboard of 25% is added to calculate the total height of bund.

4. Length of contour bund

The length of a contour bund depends on the horizontal interval of the bund. It is calculated using the following relation:

$$L = \frac{100000}{HI}$$

Using $HI = \dfrac{VI}{S} \times 100$ in above relation, we get

$$L = 100 \times \frac{S}{VI}$$

5. Area lost due to contour bunding

Area lost due to bunding is given as:

$$A_L = Length\ of\ bund\ (L) \times Bottom\ width\ of\ bund\ (B)$$

$$A_L = \frac{100000}{HI} \times B$$

Or

$$A_L = 100 \times \frac{S}{VI} \times B$$

The above equation gives the area lost due to the main bunds in the field. In addition to this, area is also lost due to the side bunds and lateral bunds which is taken as 30 percent of area lost due to main bunds. Therefore, total area lost in bunding is given as:

$$A_L = 1.3 \times 100 \times \frac{S}{VI} \times B$$

6. Earthwork

The total earthwork involved in bunding is due to the main bunds, side bunds and lateral bunds. The earthwork is simply computed using the formula for volume.

Earthwork = Cross sectional area of bund × Length of bund

If A is the cross-sectional area of bund and L is the length of bund, then Earthwork in the main bund (Em) is given as:

$$E_m = A \times L$$

Using the relation for length of the bund in above equation, we get:

$$E_m = A \times 100 \, \frac{S}{VI}$$

The earthwork of side bunds and lateral bunds is taken as 30 percent of earthwork of the main bund. Therefore, the total earthwork involved in bunding is given as

$$E_t = 1.3 \times A \times 100 \, \frac{S}{VI}$$

The cross-sectional area of bund is given as:

$$A = \frac{Top \ width + Bottom \ width}{2} \times Height \ of \ bund$$

Terracing

Terracing is another engineering or mechanical measure used for controlling soil erosion in highly sloping lands. Terraces are considered as one of the most evident anthropogenic imprints on the landscape, covering a considerable part of terrestrial landscapes (Tarolli et al. 2014). It is used extensively across diverse landscapes such as in areas where severe water erosion, mass movement and landslides from steep slopes threaten the security of land productivity, the local environment and human infrastructure (Lasanta et al. 2001). Terraced slopes even became the ideal sites for early human settlement and agricultural activities (Stanchi et al. 2012), with ancient agricultural terraces serving as pronounced evidences of ancient human history, diverse cultures and civilizations (Pietsch and Mabit 2012, Calderon et al. 2015). Terracing reduces both the length of the slope and the degree of slope. It is usually practiced in those areas where the land slope is more than 10%, rainfall is high and soils are highly erodible. It is not practiced in areas having relatively flat topography and shallow soils. Terracing has been used to conserve water, reduce erosion, expand high-quality croplands and restore degraded habitats (Bruins 2012). More recently, this practice has been found to improve other ecosystem services (ESs), such as carbon sequestration, food security as well as recreation (Ore and Bruins 2012, Garcia-Franco et al. 2014).

Terraces are classified as bench terraces and broad-based terraces. Bench terraces are step like construction formed across the land slope to intercept the runoff and minimize the soil erosion. Bench terraces break the original slope of the land and convert it into the step like fields and hence make the hilly lands suitable for cultivation. On the other hand, broad based terraces are series of broad channels and embankments constructed along the contour on the gentle slopes. These terraces are built with either a uniform or a variable but non erosive grade leading to safe disposal of the runoff. They are also known as channel terraces.

Types of bench terraces

The bench terraces are of three types:

1. Level bench terraces
2. Bench terraces sloping inwards
3. Bench terraces sloping outwards

Level bench terraces: These bench terraces are constructed in areas receiving medium rainfall and have highly permeable and deep soils. The runoff generated in these terraces is expected to get absorbed by the soils and no overflow is expected in these terraces. The level bench terraces are suitable in areas where high water requiring crops, like paddy, are cultivated. Therefore, these terraces are also known as paddy terraces or table top terraces. The slope of level bench terraces is as mild as 1 percent so as to have a proper impounding of water.

Bench terraces sloping inward: These bench terraces are constructed in areas which receive heavy rainfall and have less permeable soils. The runoff generated in these terraces is quite large, therefore a drainage channel is provided for the safe disposal of runoff towards the inner side of the terrace. Crops which are susceptible to the water logging, like potato, are cultivated on these terraces. These terraces are also known as hill type bench terraces.

Bench terraces sloping outward: These bench terraces are constructed in areas which receive medium rainfall and have permeable soils with medium depth. In order to retain the runoff within the terrace, the shoulder bund is provided on the outer end of the terrace. The shoulder bund also imparts stability to the terrace. These terraces are also known as orchard type bench terraces.

Design specifications of bench terraces

The bench terraces are designed based on the rainfall characteristics, soil type, soil depth and slope of the area. Following are the design specifications of the bench terraces:

1. Type of bench terrace

The type of bench terrace to be constructed depends on the rainfall characteristics and soil depth of the area.

2. Terrace spacing

The terrace spacing is expressed in terms of the vertical interval between the two terraces. It depends on the soil type and slope of the area. Many empirical relationships are available to compute the vertical interval between the terraces.

$$VI = 0.3 \left(\frac{S}{2} + 2 \right)$$

Where VI is vertical interval (m) and S is the land slope (percent)

There are three different cases to calculate the vertical interval:

a. When the terrace cut is vertical

$$D \text{ or } VI = \frac{WS}{100}$$

b. When the batter slope is 1:1

$$\frac{D/2}{D/2 + W/2} = \frac{S}{100}$$

Solving this, we get

$$D = \frac{WS}{100 - S}$$

c. When the batter slope is ½:1

$$\frac{D/2}{D/4 + W/2} = \frac{S}{100}$$

Solving this, we get

$$D = \frac{2WS}{200 - S}$$

3. Terrace width

The width of terrace depends on the terrace spacing and slope of the area. The width of terrace is decided on the basis of the use it is put to after the construction. The width should be such that it allows the farm operations without any hinderance. Once the width of terrace is decided, vertical interval or terrace spacing is calculated using the above equations.

4. Terrace gradient

Terrace gradient is important for the proper designing of the terraces in high rainfall areas. Proper gradients are required to safely dispose of the runoff generated on the terrace. The gradient should be such that it neither causes erosion nor water logging on the terrace. It is decided based on the maximum rainfall that has occurred in the area and peak discharge.

5. Terrace cross section

The four parameters which are taken into account to design the terrace cross section are:

a. Batter slope
b. Dimension of the shoulder
c. Inward slope of bench terrace and size of drainage channel
d. Outward slope in case of terraces sloping outwards

    a. *Batter slope*: The batter slope is provided to impart stability to the fill material. The batter slope depends primarily on the soil material used for filling. The range of batter slope varies with the soil type. Flatter (or milder) the batter slope, larger is the area lost in terracing. The vertical cuts are used in very stable soils and when the depth of the cut is not more than 1m. Batter slope of 0.5:1 is used in loose and unstable soils.

    b. *Dimension of shoulder bund*: The shoulder of larger cross section is necessarily used in terraces sloping outwards and table top terraces in order to retain the runoff water within the terrace. In the case of inward sloping terraces, the size of the shoulder bund should be nominal. The shoulder bund also provides stability to the terraces.

    c. *Inward and outward slope*: The inward and outward slope of the bench terraces depends on the type of soil and average rainfall of the area. The inward slope may vary between 2 to 10%, whereas the outward slope varies between 2 to 8%. The size of the drainage channel is calculated on the basis of runoff rate to be disposed from the terrace area. The proper grade in the range of 0.5 to 1.0% is provided in the drainage channel.

6. Earthwork

The earthwork for the bench terraces is computed using the following formula:

$$E = \frac{100WS}{8}$$

where, W is the width of terrace (m)

S is the slope (%)

In addition to above described terraces, there are some other types of terraces, which are described below:

Puerto Rico Terrace

Puerto Rico Terraces (PRT) are constructed using dry stones by excavating the soil a little during every ploughing and gradually developing the bench by pushing the soil downhill against a mechanical or vegetative barrier. These terraces are very effective in rain fed areas to increase the crop productivity. They are suitable in those areas where stones are easily available.

Stone wall terrace

Stone wall terraces are constructed where the stones are available in abundance. They are constructed across the small gullies to impound water and cause sedimentation in the upstream. They are also constructed in hilly areas in order to create the additional land for cultivation by cutting the hill slopes and concentrate the eroded soil of the adjoining lands at appropriate places (Singh 2000).

Soil Stabilizers/Additives/Conditioners

Increasing aggregate stability at the soil surface and preventing clay dispersion is known to control seal formation, increase the infiltration rate, and reduce runoff in cultivated soils. Stable aggregates at the soil surface are less susceptible to detachment by raindrop impact and to transportation by runoff water. Aggregate stability can be improved by applying soil amendments or soil stabilizers to the soil. These stabilizers include organic by-products, polyvalent salts and various synthetic polymers. High cost of soil stabilizers limits their use for agricultural purposes but can be effectively used at special sites such as at sand dunes, road cuttings, embankments and stream banks, to provide temporary stability prior to the establishment of a plant cover. Out of various soil amendments/stabilizers, gypsum and synthetic organic polymers are commonly used.

Gypsum

Gypsum is a relatively common mineral that is widely available in agricultural areas. It is mainly used as amendment for sodic soil reclamation because of its low cost, availability and ease of handling. Sodic soils are particularly susceptible to tunnel erosion. Their high Na content results in the dispersion of clay minerals and causes structural deterioration. The Ca caution present in gypsum replaces the Na^+ ion adsorbed on clay particles. But a good drainage system is required to wash out the Na from the soil. Because of low solubility of gypsum, phosphogypsum (PG) is also used because it is more rapidly soluble than mined gypsum. PG addition at the rate of 5 Mg ha^{-1} resulted in a decrease in runoff of 0.3–2.5 times and roughly decreased soil loss by 50% compared with the control. Application of gypsum enhances flocculation and result deposition of suspended clay size particles in the runoff water. It increases surface aggregate stability, thus fewer particles are detached by raindrops or overland flow.

Polymers

These soil conditioners are of two types, hydrophobic which decrease infiltration and increase runoff and hydrophilic which increase infiltration and decrease runoff. Synthetic organic polymers are more effective for a longer period as compared to natural polymers. Polyacrylamide (PAM) and polysaccharide (PSD) are two synthetic organic polymers that have recently been extensively investigated with respect to their efficacy as soil conditioners. The results of various field experiments demonstrate that spreading a small amount of PAM at the soil surface has a long-term effect on stabilizing the aggregates at the soil surface and reducing runoff, which lasts over the entire rain/irrigation season. A small amount of polymers (10–20 kg ha^{-1}) sprayed directly onto the soil surface or added to the applied water leads to stabilization and cementation of aggregates at the soil surface and hence increases their resistance to seal formation and soil erosion in the hill slopes and adjusting the eroded soil of the adjoining lands at the appropriate places.

References

Adimassu, Z., K. Mekonnen, C. Yirga and A. Kessler. 2012. Effect of soil bunds on runoff, soil and nutrient losses and crop yield in the Central Highlands of Ethiopia. Land Degrad. Dev. 25: 554–564. doi.org/10.1002/ldr.2182.

Blanco-Canqui, H. and R. Lal. 2010. Soil erosion and food security. pp. 493–512. *In*: Blanco-Canqui, H. and R. Lal (Eds.). Principles of Soil Conservation and Management. Springer, Dordrecht.

Bochet, E., J.L. Rubio and J. Poesen. 1998. Relative efficiency of three representative matorral species in reducing water erosion at the microscale in a semi-arid climate. Geomorphology 23: 139–150.

Bruins, H.J. 2012. Ancient desert agriculture in the Negev and climate-zone boundary changes during average, wet and drought years. J. Arid Environ. 86: 28–42.

Calderon, M.M., N.C. Bantayan, J.T. Dizon, A.J.U. Sajise, A.L. Codilan and M.S. Canceran. 2015. Community-based resource assessment and management planning for the rice terraces of Hungduan, Ifugao, Philippines. J. Environ. Sci. Manag. 18: 47–53.

Dass, A., S. Sudhishri, N.K. Lenka and U.S. Patnaik. 2011. Runoff capture through vegetative barriers and planting methodologies to reduce erosion, and improve soil moisture, fertility and crop productivity in southern Orissa, India. Nutri. Cycl. Agroecosyst. 89: 45–57.

de Baets, S., J. Poesen, A. Knape, G.G. Barbera and J.A. Navarro. 2007. Root characteristics of representative Mediterranean plant species and their erosion-reducing potential during concentrated runoff. Plant Soil 294: 169–183.

Dimelu, M.U., S.E. Ogbonna and I.A. Enwelu. 2013. Soil Conservation practices among arable farmers in Enugu-North Agricultural Zone, Nigeria: Implication for climate change. J. Agric. Ext. 17: 184–196.

Durán, Z.V.H., M.J.R. Francia, P.C.R. Rodríguez, R.A. Martínez and R.B. Cárceles. 2006. Soil erosion and runoff prevention by plant covers in a mountainous area (SE Spain): implications for sustainable agriculture. Environmentalist 26: 309–319.

Edward, L.S. and J.D Simon. 2003. Soil erosion and conservation pp. 23–30. *In:* Nieder, R. and D. Benbi. Handbook of Processes and Modelling in the Soil-Plant System. Ist Edition. CRC Press. Boca Raton, USA.

ELD. 2015. The value of land: Prosperous lands and positive rewards through sustainable land management. The Economics of Land Degradation. Available from www.eld-initiative.org.

FAO and ITPS. 2015. The Status of the World's Soil Resources (Main Report). Food and Agriculture Organization of the United Nations, Rome.

Garcia-Franco, N., M. Wiesmeier, M. Goberna, M. Martinez-Mena and J. Albaladejo. 2014. Carbon dynamics after afforestation of semiarid shrublands: implications of site preparation techniques. For. Ecol. Manag. 319: 107–115.

Gebreegziabher, T., J. Nyssen, B. Govaerts, F. Getnet, M. Behailu, M. Haile and J. Deckers. 2009. Contour furrows for *in situ* soil and water conservation, Tigray, Northern Ethiopia. Soil Tillage Res. 103: 257–264. DOI: 10.1016/j.still.2008.05.021.

GSP. 2017. Global Soil Partnership Endorses Guidelines on Sustainable Soil Management http://www.fao.org/global-soil-partnership/resources/highlights/detail/en/c/416516/.

Kato, E., C. Ringler, M. Yesuf and E. Bryan. 2011. Soil and water conservation technologies: a buffer against production risk in the face of climate change? Insights from the Nile basin in Ethiopia. Agric. Econ. 42: 593–604.

Kell, W.V. and G.F. Brown. 1937. Strip cropping for soil conservation. Washington, D.C. U.S. Dept. of Agriculture. USA.

Lasanta, T., J. Arnaez, M. Oserin and L.M. Ortigosa. 2001. Marginal lands and erosion in terraced fields in the Mediterranean mountains: a case study in the Camero Viejo (Northwestern Iberian System, Spain). Mt. Res. Dev. 21: 69–76.

Liu, Q.J., H.Y. Zhang, J. An and Y.Z. Wu. 2014. Soil erosion processes on row sideslopes within contour ridging systems. Catena 115: 11–18.

Ma, Q., X. Yu, G. Lu and Q. Liu. 2010. Comparative study on dissolved N and P loss and eutrophication risk in runoff water in contour and down-slope. J. Food Agric. Environ. 8: 1042–1048.

Martínez, R.A., Z.V.H. Durán and F.R. Francia. 2006. Soil erosion and runoff response to plant cover strips on semiarid slopes (SE Spain), Land Degrad. Dev. 17: 1–11.

Mohamed, H.H. 2015. Cause and effect of soil erosion in Boqol-Jire Hargeisa, Somaliland. Ph.D. Thesis, University of Hargeisa, Somalia.

Obert, J., L. Paramu, C. Mafongoya, M. Chipo and M. Owen. 2016. Seasonal Climate prediction and adaptation using indigenous knowledge systems in agriculture systems in Southern Africa: A Review. J. Agril. Sci. 2: 23–27.

Oldeman, L.R. 1992. Global extent of soil degradation. ISRIC Bi-Annual Report. Wageningen, The Netherlands. pp. 19–36.

Ore, G. and H.J. Bruins. 2012. Design features of ancient agricultural terrace walls in the Negev desert: human-made geodiversity. Land Degrad. Dev. 23: 409–418.

Pietsch, D. and L. Mabit. 2012. Terrace soils in the Yemen Highlands: Using physical, chemical and radiometric data to assess their suitability for agriculture and their vulnerability to degradation. Geoderma 185-186: 48–60.

Puigdefabregas, J. 2005. The role of vegetation patterns in structuring runoff and sediment fluxes in drylands. Earth Surf. Process. Landf. 30: 133–147.

Raes, D., E.M. Kafiriti, J. Wellens, J. Deckers, A. Maertens, S. Mugogo, S. Dondeyne and K. Descheemaeker. 2006. Can soil bunds increase the production of rain-fed lowland rice in south eastern Tanzania? Agric. Water Manage. 89: 229–235.

Ramajayam, D., P.K. Mishra, S.K. Nalatwadmath, B. Mondal, R.N. Adhikari and B.K.N. Murthy. 2007. Evaluation of different grass species for growth performance and soil conservation in Vertisols of Karnataka. Indian J. Soil Conserv. 35: 54–57.

Rey, F. 2003. Influence of vegetation distribution on sediment yield in forested marly gullies. Catena 50: 549–562.

Roose, E. 1992. Contraintes et espoirs de development d'ine agriculture durable en montages tropicales. Bull. Reseau. Erosion. 12: 57–70.

Singh, M.J., A. Yousuf, S.C. Sharma, S.S. Bawa, A. Khokhar, V. Sharma, V. Kumar, S. Singh and S. Singh. 2017. Evaluation of vegetative barriers for runoff, soil loss and crop productivity in Kandi region of Punjab. J. Soil Water Conserv. 16: 325–332.

Singh, P.K. 2000. Watershed Management (Design and Practices). Agarwal Printers Private Limited, Udaipur, Rajasthan. India.

Stanchi, S., M. Freppaz, A. Agnelli, T. Reinsch and E. Zanini. 2012. Properties, best manage-ment practices and conservation of terraced soils in Southern Europe (from Mediter-ranean areas to the Alps): a review. Quat. Int. 265: 90–100.

Suresh, R. 2016. Soil and Water Conservation Engineering, Standard Publishers Distributors, New Delhi, India.

Tarolli, P., F. Preti and N. Romano. 2014. Terraced landscapes: from an old best practice to a potential hazard for soil degradation due to land abandonment. Anthropocene 6: 10–25.

UNCCD. 2015. Desertification, Land Degradation & Drought (DLDD): some global facts and figures. United Nations Conventions to Combat Desertification. Available from: http://www.unccd. int/Lists/SiteDocumentLibrary/WDCD/DLDD%20Facts.pdf.

Vashisht, B.B., B.S. Sidhu, S. Singh and N. Bilwalkar. 2013. Effect of different mulches on soil erosion and carry-over of residual soil moisture for sowing of crop in maize-wheat cropping sequence in rainfed Shivaliks of Punjab. Indian J. Soil Conserv. 41: 136–140.

Wainwright, J., A.J. Parsons and W.H. Schlesinger. 2002 Hydrology–vegetation interactions in areas of discontinuous flow on a semi-arid bajada, Southern New Mexico. J. Arid Environ. 51: 319–338.

Chapter 6

Gully Erosion and its Control

Mahesh Chand Singh

Introduction

The advanced stage of rill or channel erosion is termed as gully erosion, which cannot be smoothened by ordinary tillage practices. The gully development process follows sheet and rill erosions, thereby resulting in the removal of soil along drainage lines by surface runoff water. In addition to the natural depressions on the land surface responsible for runoff accumulation, the unchecked rills may also be encouraging the process of gully erosion. Once started, gullies will continue to move by headword erosion or by the slumping of the side walls unless steps are taken to stabilize the disturbance (Suresh 2018). With the advancement in gully development through accelerated water erosion, sediment transport gets significantly enhanced. The gullies formed are mainly either U-shaped or V-shaped channels of at least 30 cm wide or 30 cm deep.

The rate of gully erosion is predominantly dependent on the runoff-producing features such as drainage area, soil characteristics, shape, size and alignment of gully, and the slope of the watershed channel. Moreover, the rate and the extent of gully development is directly associated with the volume and velocity of runoff water. The higher volume of runoff water in the absence of natural or perennial vegetation inclines to detach and transport a relatively larger volume of soil mass. Globally, the soil and water loss due to gully erosion has become one of the foremost factors limiting local economic development. In other terms, gully erosion has been considered as an important environmental hazard throughout the

Punjab Agricultural University, Ludhiana.
Email msrawat@pau.edu

world which affects several soil and land functions (Ionita et al. 2015). The factors responsible for increasing the rate of soil erosion also include land topography including slope, vegetative cover of soil surface, characteristics of rainfall and the resistance offered by top soil and underlying hard layer.

At present, gully erosion has emerged out to be a major challenge in the world to impart a negative impact on agricultural production, land value, infrastructures, landscapes, arable farmlands and vegetation, and soil fertility or productivity as well as human and animal lives (Poesen 2011, Abdulfatai et al. 2014). Gully erosion results in a significant loss of soil, roads and bridges and land productivity. It also reduces water quality through an increased sediment load in the streams. The soil type also influences the gully erosion to a great extent. According to Poesen et al. (2003), gully-based valley sediment yield contributes 10 to 94% of the total watershed sediment yield. A well-developed soil (e.g., oxisol) is more homogeneous, cohesive and well formed with lateritic behaviour, thereby increasing its resistance to erosion. However, the granite-gneiss saprolite structure is less cohesive having reduced resistance to erosion due to the presence of weathered minerals and other elements such as kaolinite (de Freitas Sampaio et al. 2016).

Stages of Gully Development

There are four stages of gully development:

Stage–1: Formation stage

This is the initial stage of gully formation which usually proceeds gradually. In this stage, channel erosion and the deepening of the gully bed takes place.

Stage–2: Development stage

In this stage of gully development, the gully depth and width gets inflated due to runoff received from the up-stream part of the gully head.

Stage–3: Healing stage

In this stage, vegetation starts growing in the channel and no considerable erosion occurs from the gully section.

Stage–4: Stabilization stage

In this stage, the gully becomes fully stabilized and the gully development stops unless the healing process is disturbed. The channel and gully walls achieve a stable gradient and slope respectively with the growth of vegetation cover in abundance on soil surface.

Classification of a Gully

A gully can be classified as follows:

- Based on gully depth and drainage area
- Based on depth, width and slope of the gully
- Based on shape of gully cross-section
- Based on state of the gully
- Based on continuity of gully

Based on gully depth and drainage area

Small gully

A gully having depth of less than 1 m and drainage area less than 2 ha is termed as a small gully. A small gully can be easily crossed by farm implements and smoothened by ploughing or any other land development operations. It can also be stabilized through vegetation.

Medium Gully

A gully having depth and drainage area in the range of 1–5 m and 2–20 ha respectively, is termed as a medium gully. A medium gully cannot be crossed easily by farm implements and can be stabilized through tillage operations or terracing. The sides of the gully can be stabilized through vegetation.

Large gully

A gully having a depth more than 5 m and drainage more than 20 ha is termed as a large gully. A large gully cannot be reclaimed and tree plantation can be adopted as an effective control measure.

Based on depth, width and slope of the gully

In this type of classification, gully formation is categorized into four groups viz. G–1, G–2, G–3 and G–4. This classification is based on the depth, width and slope of the gully (Table 1).

Table 1. Gully types based on depth, width and slope.

Parameter	Gully type			
	G–1	G–2	G–3	G–4
Depth (m)	≤ 1.0	1.0–3.0	3.0–9.0	> 9.0
Width (m)	< 18.0	< 18.0	18.0	> 18.0
Side slope (%)	< 6.0	< 6.0	6.0–12.0	> 12.0

Based on shape of gully cross-section

- U-Shaped
- V-Shaped
- Trapezoidal

U-Shaped gullies

- These are formed where both the surface and sub-surface soil have the same resistance against erosion. These types of gullies are found in alluvial plains.
- Features of U-shaped gullies:
  - They have a U-shaped cross-section
  - They have lower flow velocity compared to V-shaped gullies
  - They carry a massive discharge which is contributed from a large catchment area
  - The longitudinal slope of the gully bottom and slope of the land through which the gully passes normally remain parallel
  - Grow wider and longer, but not deeper
  - Continue to grow headword
  - Active erosion takes place from the side banks and the gully head due to undercutting at the base of the vertical cut
  - They have a large lateral spacing

V-Shaped gullies

- This is the most common form of gully. These are formed where the sub-surface soil has more resistance against erosion than surface soils. Their shape is dependent upon the soil features, age of the gully, kinds of erosion and climate of the area under consideration.
- Features of V-shaped gullies:
  - They are V-shaped in cross-section
  - Can carry a smaller discharge through them but have a higher velocity
  - Frequently developed from rill erosion when water from several rills contributes into a single rill
  - They have a smaller lateral spacing between the gullies
  - Largely appear at a steep slope
  - The longitudinal gradient of the channel is greater than the land slope
  - Carry runoff from a relatively small catchment area
  - V-shaped gullies create difficulty in contour cultivation

Trapezoidal shaped gullies

- Such gullies are formed where the bottom of the gully is made of more resistant material than the surface soil. The further development or advancement of such gullies is not possible.

Based on state of the gully

- Active gullies
- Inactive gullies

Active gullies

The gullies found in plain areas are active gullies and their dimensions are enlarged with the passage of time. The enlargement of gully size is dependent on the soil features, land use and runoff volume passing through it.

Inactive gullies

They are found in rocky areas and their dimensions do not change considerably with the passage of time due to higher resistance by rocks to erosion through runoff water.

Based on continuity of gully

- Continuous gully
- Discontinuous gully

Continuous gully

A continuous gully has a main gully channel and is comprised of several mature or immature branch gullies and many branch gullies. A gully system (gully network) is comprised of several continuous gullies.

Discontinuous gully

A discontinuous gully is also termed as an independent gully and can develop on hillsides after land sliding. Initially, a discontinuous gully does not have a distinct junction with the stream channel or main gully and the water spreads over a closely flat area. However, with passage of time, it reaches the stream or main gully channel.

Factors responsible for acceleration of gully erosion

- Improper land use, such as compacted roads without a proper drainage system (de Freitas Sampaio et al. 2016)
- Expansion in rural and urban development
- Diverting the runoff water towards more erodible and fragile soil
- Removal of resistant soil and exposing of the susceptible one

Causes of Gully Formation

Gully formation is caused by both natural and anthropogenic interventions.

Natural or physical factors

- Precipitation (monthly rainfall distribution), rainfall intensity, runoff and rapid snowmelts
- Increased runoff triggered by low levels of vegetation cover and/or poor infiltration rate of soils
- Flooding
- Soil properties–soil texture
- Topography-shape of catchment, size of catchment, and length and gradient of the slope
- Poor vegetative cover due to overgrazing, fires or salinity problems
- Length and gradient of slope

Man-made factors

- Improper land use, e.g., improper design, construction and maintenance of waterways in cropped areas
- Diversion of a drainage line to an area of high risk to erosion
- Forest and grass fires
- Livestock and vehicle trails
- Mining
- Destructive logging
- Overgrazing, e.g., grazing on soils susceptible to gully erosion
- Road construction

Adverse Impacts of Gully Erosion

- Reduction in available land for agriculture (Zgłobicki et al. 2015)
- Decrease in crop yields (Zgłobicki et al. 2015)
- Land productivity reduces significantly due to loss of sediment from the valley (Zgłobicki et al. 2015, Zgłobicki et al. 2015)
- Increased flooding and sediment generation negatively affect the farms, fences, roads, railways, bridges and culverts
- A continuous change in local groundwater levels and the vegetation near the gully
- Negative impact on building, farmlands and other properties as well as human and animal life (Danladi and Ray 2014, Ezeigwe 2015, Angela and Ezeomedo 2018)
- Augmented turbidity and nutrient loads in ponds, streams, rivers, dams and running water

- Ecological problems including damage to aquatic habitation
- The access to affected land becomes very difficult
- Destruction of underground utilities such as phone cables, power and other pipes
- Higher possibility of creek erosion within downstream waterways as a result of increased sediment flow
- Issues of water quality within the downstream dams and waterways in relation to released sediments containing nutrients and metals
- Possible discharge of salts from soil and runoff acidic in nature into receiving waters
- Entry of sediments into waterways and water supplies
- Problem of sedimentation within the downstream waterways
- The reintegration of downstream water bodies through de-silting may involve huge costs
- Loss of storage capacity of reservoirs
- Recreational impacts
- Adverse influence on other water resources

Gully Control Measures

The gully control includes the following:

- Regulation and reduction of run-off rates through improvements in the gully catchments
- Diversion of surface runoff above the gully area
- Gully stabilization through structural measures and the accompanying re-vegetation

Gully Control by Vegetation

Vegetative cover protects the gully against scouring by reducing the flow velocity of water through an increase in hydraulic resistance of the channel section. Vegetation is considered as the foremost, long-term defence for controlling gully erosion. However, structures might be required to stabilise a gully and prevent from further erosion. Structures are either temporary or permanent. Concrete, masonry, wood or other materials can be used for building the permanent structures. Permanent structures require technical skills for their design and construction.

Gully control by structures:

Following two types of structures are used for controlling gully erosion.

- Temporary structures
- Permanent structures

Temporary structures are used to encourage the vegetative growth on the upstream portion of the gully by collecting a sufficient quantity of soil. Secondly, to check the status of gully erosion untill plentiful vegetation is established at the critical points of the gully. The various structures (temporary and permanent) used for gully control are listed in Table 2.

Table 2. Examples of temporary and permanent gully control structures.

Temporary structure	Permanent structure
Brushwood dam i. Single row brushwood dam ii. Double row brushwood dam	Spillway – Drop inlet spillway – Chute spillway – Straight drop spillway or Drop spillway or Drop structure
Loose rock dams and rock filled dams	Concrete dam
Log check dam	Masonry check dam
Woven wire check dam	Gabions check dam
Netting dam	
Staggered trenches or bunds	
Terraces	
Grassed waterways	

A few commonly used temporary and permanent structures of gully control are discussed below:

Temporary structures

The temporary structures viz. woven wire dam, brushwood dam and loose rock dams are practiced in G-1 type gullies (Table 2), in order to keep the gullies stable and aid the establishment of vegetative cover. They are simple in construction and maintenance, and can be made from easily available local materials. These structures can have a life span of about 3 to 8 years.

Woven wire dams

They are used in gullies having a moderate slope with a small drainage area for control of erosion through the establishment of vegetation. For building a woven wire dam, a row of posts at about 1.2 m intervals and 60–90 cm deep is set along the curve of the proposed dam. A heavy gauge woven wire is placed against the post with the lower part set in a trench of 15–20 cm deep projecting 25–30 cm above the ground surface along the spillway interval. Rock, sod or brush can be placed at a length of 1.2 m approximately to form the apron. Straw, fine brush or similar materials would be positioned against the wire on the upstream side to the height of spillway crest for sealing the structure.

Brush wood dams

They are the least stable among all types of check dams and best suited for gullies having a smaller drainage area. Brushwood dams are of two types viz. single post row brushwood check dams and double post row brushwood check dams. The double post row brushwood check dams are used for handling high runoff. The central part of the dam is kept lower than at the ends. The bottom and sides of the gully are covered with straw or similar fine mulch in a thin layer for about 3–4.5 m along the site of the structure. Brushes are packed closely above the mulch at about one half of the dam height. Several rows of stakes are driven crosswise in the gully having a spacing of 60 cm and 30–60 cm from row to row and stake to stake respectively. For fastening the stakes in a row, heavy galvanized wire is used. Large stones can also be placed on the top of bushes to keep it compressed and in close contact with the gully bottom.

Loose rock dam

These are the most suitable for gullies having a small to medium drainage area. Loose rock dams and rock filled dams are used in an area where rocks or stones of good quality in the appropriate sizes (e.g., flat stones) are available. For building a loose rock dam, a trench is made across the gully to a depth of 30 cm in which the stones are then laid in rows to a desired height. The centre of the structure is kept lower compared to the sides to form a spillway.

Design criteria for temporary structures

- The overall height of the temporary check dam should be limited to 75 cm with an effective height of 30 cm. A minimum freeboard of 15 cm is essential.
- Their lifespan should be 3 to 8 years thereby having been designed for a rainfall return period of 10 years.
- Check dams tentatively should be placed in such a manner that the crest elevation of one will be same as the bottom elevation of the nearby dam upstream. The check dam of a lesser height with a higher slope will need more recurrent check dams down the stream.
- An apron should be provided to prevent the scouring due to the flow passing over the check dams. For this purpose, rip-rap is provided at the length of 1 to 1.5 m downstream of the check dams.

Permanent structures

Straight drop spillway

It is one of the highly suitable permanent gully control structures mainly used at the gully bed to create a control point. Many such drop structures are built across the gully width at fixed intervals throughout the length to develop a continuous break to the water flow for the deposition of sediments and the filling of the gully section. Drop structures may also be used at the gully head for a risk free flow and controlling of the gully head. A free board of 15–30 cm is required. The components of a drop structure include head wall, head wall extension, side walls, wing walls, apron, longitudinal sills, end sill and cut-off walls. This structure is recommended in the G-2 type of gullies where the depth is limited to 3 m. A drop structure has the following three major purposes.

- To provide a transition between a waterway (broad or flat) and a ditch or gully section
- To raise the flow line of the waterway in order to provide drainage in the case of wet waterways
- To form a sufficient soil depth for vegetative growth, where the gully bottom is found to be at risk

Chute spillway

It is constructed on a steep slope with a suitable inlet and outlet used at the locations where the head drop varies from 5 to 6 m. It is usually built to handle full flow at the gully head. The chute spillway handles the flow with a super critical velocity. This structure is recommended for G-3 type gullies where the depth is more than 3 m.

The chute spillways are suitable for the following conditions:

- It can be used at the sites where the conditions are not suitable for constructing the check dams
- It can be used suitably in combination with different structures such as check dams and other detention-type structures
- For high over falls where a full flow structure is required

Drop inlet spillway

This structure consists of a conduit connected with a suitable inlet and outlet. The inlet is built with a drop to guide the water flow from inlet to conduit (e.g., hood type). The outlet generally consists of a chute or a propped pipe with a suitable size to allow an adequate flow to the downstream channel. This structure is recommended in the G-4 type of gullies where depth is more than 9 m and has a steep slope. The functions of different components of drop structure are listed in Table 3.

The uses of different permanent spillway structures are listed in Table 4, whereas, the advantages and limitations are listed in Table 5. The different methods for computing the maximum discharge through a gully are listed in Table 6.

Table 3. Functions of different components of a drop structure.

Sr. No.	Component	Function
1	Head wall	– Easy conveyance of water – Acts as a front wall against flow of runoff in the drop spillway
2	Head wall extension	– To check the flow of water from the sides of the structure – To provide structural strength against sliding of the structure
3	Side walls	– To prevent splashing of water flow over the gully banks – To confine the water flow within the apron
4	Wing walls	– To prevent the flow backward into the space left between gully wall and side wall of the structure
5	Apron	– To dissipate the maximum kinetic energy of falling water by creating hydraulic jump – To reduction the velocity of outgoing water
6	Longitudinal sills	– For making the apron stable
7	End sill	– For blocking the water going directly into the channel below
8	Cut-off walls	– For offering structural strength of the structure against sliding

Table 4. Uses of different spillway structures.

	Drop inlet spillway
1	As a principal spillway in debris basins
2	As a principal spillway in reservoirs or farm ponds for letting out the harvested water
3	As a water inlet channel in drainage or irrigation structures
4	As a culvert in forest roads
5	For controlling the flood
6	For safe disposal of water harvested water in conjunction with check dams
7	Stabilization of gully grade
	Chute spillway
1	For conservation of water and collection of sediments
2	To control the advancement in gully head
3	To safely convey the runoff from upstream areas into the gully without erosion
4	To control gradient of artificial or natural channels
	Straight Drop Spillway or drop structure
1	As an outlet in the tile drainage system
2	For controlling erosion in order to protect the building, roads, etc.
3	For stabilization of the grade in the lower reaches of the outlets and waterways
4	It can be used in the water distribution system for controlling irrigation
5	For letting out the water from reservoirs
6	For controlling the tail water at the outlet section of the conduit
7	For release of irrigation water into the field

Table 5. Advantages and limitation of different spillway structures.

	Advantages	Limitations
	Drop inlet spillway	
1	One of the most efficient structures for the stabilisation of gully grade and prevention of flood	More susceptibility to get chocked by presence of debris in the water
2	Less construction materials compared to the straight drop spillway for the same drop	It is not suitable for use at places where greater earthwork is vigorous for construction
3	Lesser construction cost	Spillway capacity can be reduced
	Chute spillway	
1	Comparatively, are easy to construct	Problem of seepage in poorly drained areas
2	Very stable having lower chances of serious structural damage compared to other types of structures	Risk of undermining due to rodents
3	Lesser chances to be clogged by debris compared other structures in relation to their discharge capacities	It requires extra effort such as thorough compaction of the construction site in terms of time and money
	Straight Drop Spillway or drop structure	
1	It is easy to construct	A stable grade of gully is required for construction
2	The risk of undermining by rodents is not possible.	The construction becomes costly in gullies, where discharge < 3 m³/s and the drop or total head > 3 m
3	The clogging of the conduit by debris is not a problem	Technically, the construction of this structure cannot be justified particularly for temporary storage
4	Lower susceptibility to structural damage compared to other structures	

Concrete dam

If the materials for construction of masonry check dams are not sufficient, concrete dams are recommended, suitably using the specifications given for the construction of masonry dams. A concrete dam if damaged cannot be repaired easily. For constructing a concrete dam, cement of good grade, reinforcement and steel bars are needed. The buttresses are made to support the head wall, particularly if the length of spillway is more than 3 m. For protecting the structure from sliding, the cut-off walls should be placed at fairly greater depth both in the gully bed and sides.

Rubble masonry dam

In gullies, where the runoff of rates is very high and vegetation cannot be established, these dams can be used. Their construction is recommended where stones or rocks are easily available in adjoining areas. They are constructed with

Table 6. Computation of maximum discharge of gully catchment.

Name of method	Formula	Description
Rational formula	$$Q_{max} = \frac{CIA}{3.6}$$ This formula gives best results for torrents with catchment areas > 300 ha	– Q_{max}=Maximum discharge of the gully catchment gully catchment at the proposed check dam point (m³/s) – C=Runoff coefficient, varies from 0.20 to 0.50 depending on type of land use and topography – I=Rainfall intensity (mm/hr) – Rainfall intensity is computed based on one hour maximum rainfall intensity with a frequency of 5 to 10 years – A = Catchment area of the gully above the proposed check dam (km²)
Kresnik formulas (main)	$$Q_{max} = \frac{c \times 32 \times A}{0.5 + A^{1/2}}$$	– Q_{max}=Maximum discharge of the gully catchment gully catchment at the proposed check dam point (m³/s) – c=Coefficient, varies from 0.6 to 2.0 depending on land use type – A=Gully catchment area above the proposed check dam (km²)
Kresnik formulas (simple)	$$Q_{max} = 25 \times A^{1/2}$$ This formula gives more suitable results for gullies having catchment areas < 20 ha	
General run-off formula	$$Q_{max} = A_w \times V$$ $$V = \frac{1.486}{n} \times R^{2/3} \times S^{1/2}$$ $$R = A_w / P_w$$ For computing R, the cross-sectional area of the gully should be measured at highest flood water level	– Q_{max}=Maximum discharge of the gully catchment gully catchment at the proposed check dam point (m³/s) – A_w=Wetted or cross-sectional area of main gully bed (m²) – V=Manning's flow velocity (m/s) – n=Roughness coefficient – R=Hydraulic radius (m) – S=Slope or gradient o£ the gully channel (%) – A_w=Wetted area (m²) – P_w=Wetted perimeter (m)
Spillway formula	$$Q_{max} = C^* LD^{3/2}$$ For check dams, loose rock and boulder log, C^*=3.0 For cement masonry check dams and gabion, C^*=1.8	– Q_{max}=Maximum discharge of the gully catchment gully catchment at the proposed check dam point (m³/s) – C^*=Coefficient – L=Spillway length (m) – CLD=Spillway depth (m)

a minimum thickness of walls as 30 cm keeping the downstream slope of the dam at least 1:2 below the spillway. The thickness of the base is kept greater than or equal to 3/4th of the height of the dam. The minimum thickness of cut off walls, side walls and the apron should be about 30 cm. The thickness of the main wall from the crest of the spillway to the top of the dam should be greater than or equal to 35 cm. Up-stream side of the dam should be maintained at an angle of about

10° with the vertical for ensuring proper settling. The length of the apron should be greater than or equal to 1.5 times the dam height measured from apron floor to the spillway's crest. Drains or weep holes should also be provided near the base of the dam for drainage.

Gabion structure

It is a stone filled rectangular wire mesh box where the size of stones filled should always be greater than the mesh openings. They are flexible, permeable and economical and constructed where stones are easily available in abundance. Galvanized wires are used for making boxes to ensure longer service life and prevent from rust formation. This structure is permanent, easy to construct, efficient and economical.

Design requirements of permanent gully control structures

Gully control structures are installed across active gullies to stabilize them by controlling erosion. These structures are mainly designed for the safe disposal of surplus runoff generated from the watershed by considering the following points.

- Keeping the hydrologic design in mind, there should be a provision for the safe discharge of the runoff
- Keeping structural design in mind, there should be enough strength of the structure to endure the pressure exerted by the runoff water
- Keeping hydraulic design in mind, the structure should be protected from erosion due to the runoff passing over it.

Different methods or models used in the past to study gully erosion

- Aerial photographs and traditional photogrammetry (Gomez-Gutierrez et al. 2018)
- Global Navigation Satellite Systems (Gomez-Gutierrez et al. 2018)
- Pin networks
- Light Detection and Ranging sensors (Gomez-Gutierrez et al. 2018)
- Topographic profilers (Gomez-Gutierrez et al. 2018)
- Total stations
- Terrestrial Laser Scanners
- Terrestrial Structure from Motion (SfM)
- Ground manual measuring method (Vandaele and Poesen 1995, Casali et al. 2006)
- Field marks (Ionita 2006)
- Aerial photos and digital elevation model (Martınez-Casasnovas 2003)
- RTK-GPS measurement (Wu and Cheng 2005)

- Conventional gully delineation
- Gully morphology

Recent techniques involving remote sensing and GIS

- 3D photo-reconstruction technique
- Ground based LiDAR
- Laser profilometer
- Total station survey

Remote sensing has become a widely used technology for studying gully erosion due the multiple advantages offered by it (Wang et al. 2014). The advantages offered include the following:

- Wide detection range
- Fast acquisition of information
- Short period
- Less restrictions by ground conditions
- Rich electromagnetic wave information

Factors to be considered for controlling gully erosion

- The cause of action to be taken for controlling gully erosion
- The effect if no action is taken for controlling the gully erosion
- Catchment size
- Type of soil
- Identification of the most actively eroding gully components such as gully head, floor, sides and height
- Potential for diversion of run-off flowing into the gully to a safe area

Other factors to be considered while fixing gully erosion

- Check whether the erosion is active by looking at the gully head, walls and floor
- Check water source
- Improve groundcover
- Review land management
- Install earthworks

Suggestions for controlling the gully formation

- Gully control by constructing dams to cut down the velocity of water flow and escalation of sedimentation

- Land use correction based on natural ability and limitations in relation to geomorphologic and physiographic features of soil of a particular area (Arabameri et al. 2018).
- Curtailing the tree clearing
- Restoration of vegetation easy-going with the natural conditions of the area (Brooks et al. 2016, Arabameri et al. 2018)
- Adoption of appropriate cultivation methods
- Improved land use
- Appropriate drainage system for urban and rural areas (de Freitas Sampaio et al. 2016)
- Ensuring that the drainage from buildings, roads and stock routes is not concentrated into gullies
- Diversion of runoff from the erodible soil to safe area
- Fencing of areas susceptible to erosion
- Maintaining contour banks and waterways
- Adoption of conservation practices to prevent and mitigate the problem of gully erosion in rural areas
- In extreme cases, planting vegetation, fencing, diversion banks and engineering structures may be required to control gully erosion.
- To make the farmers aware of the region through environmental officials (Arabameri et al. 2018) about the followings:
  o The type of cultivation
  o Principles of appropriate cultivation
  o Avoidance of overgrazing and demolition of natural vegetation

Maintaining a good vegetative cover on the ground may serve as the best alternative in preventing gully erosion. Thus, maintaining a minimum 70% ground cover, a stubble cover of 30% on cultivated areas and stabilizing the points of sharp slopes and gully head may help to minimise the gully erosion to a great extent.

References

Abdulfatai, I.A., I.A. Okinlola, W.G. Akande, L.D. Momoh and K.O. Ibrahim. 2014. Review of gully erosion in Nigeria: causes, impacts and possible solutions. J. Geosci. Geomatics 2: 125–129.

Angela, A. and I.C. Ezeomedo. 2018. Changing Climate and the Effect of Gully Erosion on Akpo Community Farmers in Anambra State, Nigeria. J. Ecol. Nat. Resour. 2: 000147. doi: 10.23880/jenr-16000147.

Arabameri, A., B. Pradhan, H.R. Pourghasemi, K. Rezaei and N. Kerle. 2018. Spatial modelling of gully erosion using GIS and R programing: A comparison among Three Data Mining Algorithms. Appl. Sci. 8: 1369. doi:10.3390/app8081369.

Brooks, A., T. Pietsch, R. Thwaites, R. Loch, H. Pringle, S. Eccles, T. Baumgartl, J. Biala, J. Spencer, P. Zund, T. Spedding, A. Heap, D. Burrows, R. Andrewartha, A. Freeman, S. Lacey, W. Higham and M. Goddard. 2016. Communique: Alluvial Gully Systems Erosion Control & Rehabilitation Workshop, Collinsville 8–10 August 2016. Report to the National Environmental Science Programme. Reef and Rainforest Research Centre Limited, Cairns (23pp.).

Casali, J., J. Loizu, M.A. Campo, L.M. de Santisteban and J. Alvarez-Mozos. 2006. Accuracy of methods for field assessment of rill and ephemeral gully erosion. Catena 67(2): 128–138.

Danladi, A. and H.H Ray. 2014. Socio-economic effect of gully erosion on land use in Gombe Metropolis, Gombe State, Nigeria. J. Geogr. Reg. Plann. 7(5): 97–105.

de Freitas Sampaio, L., M.P. Pires-de Oliveira, R. Cassaro, V.G. Silvestre-Rodrigues, O.J. Pejon, J. Barbujiani-Sígolo and V. Martins-Ferreira 2016. Gully erosion, land uses, water and soil dynamics: a case study of Nazareno (Minas Gerais, Brazil). DYNA 83(199): 198–206. doi: http://dx.doi.org/10.15446/dyna.v83n199.54843.

Ezeigwe, P.C. 2015. Evaluation of socio-economic impacts of gully erosion in Nkpor and Obosi. Environ. Res. 7(7): 34–38.

Gomez-Gutierrez. A., S. Schnabel, F. Lavado-Contador, J.J. de Sanjose-Blasco, A.D.J. Atkinson, M. Pulido-Fernandez, M. Sanchez-Fernandez and A. Alfonso-Torreno. 2018. Studying gully erosion processes in rangelands of SW Spain and guiding restoration strategies using the UAV+SfM workflow. Geophy. Res. Abst. 20, EGU2018–19157–1.

Ionita, I. 2006. Gully development in the Moldavian Plateau of Romania. Catena 68(2-3): 133–140.

Ionita, I., M.A. Fullen, W. Zgłobicki and J. Poesen. 2015. Gully erosion as a natural and human induced hazard. Nat. Hazards 79: S1–S5. Doi:10.1007/s11069-015-1935-z.

Martınez-Casasnovas, J.A. 2003. A spatial information technology approach for the mapping and quantification of gully erosion. Catena 50(2-4): 293–308.

Poesen, J., J. Nachtergaele, G. Verstraeten and C. Valentin. 2003. Gully erosion and environmental change: importance and research needs. Catena 50(2-4): 91–133.

Poesen, J. 2011. Challenges in gully erosion research. Landform analysis 17: 5–9.

Suresh, R. 2018. Soil and Water Conservation Engineering, 5th edition, Standard Publisher Distributors, Delhi, India.

Vandaele, K. and J. Poesen. 1995. Spatial and temporal patterns of soil erosion rates in an agricultural catchment, central Belgium. Catena 25(1-4): 213–226.

Wang, T., F. He, A. Zhang, L. Gu, Y. Wen, W. Jiang and H. Shao. 2014. A quantitative study of gully erosion based on Object-Oriented Analysis Techniques: A Case Study in Beiyanzikou Catchment of Qixia, Shandong, China. The Sci. World J. Article ID 417325, 11pages. http://dx.doi.org/10.1155/2014/417325.

Wu, Y. and H. Cheng. 2005. Monitoring of gully erosion on the Loess Plateau of China using a global positioning system. Catena 63(2-3): 154–166.

Zgłobicki, W., B. Baran-Zgłobicka, L. Gawrysiak and M. Telecka. 2015. The impact of permanent gullies on present-day land use and agriculture in loess areas (E. Poland). Catena 126: 28–36.

Chapter 7

Soil Erosion by Water-Model Concepts and Application

Jürgen Schmidt[1], and *Michael von Werner*[2]

Introduction

The extent of soil erosion is largely determined by individual, extreme heavy rainfall events. Erosion is therefore not a continuous process, but the result of isolated individual events which cannot be directly compared with one another due to the large number of influencing variables varying over time. Even with the greatest possible effort, therefore, only individual states defined by the local conditions and the respective external circumstances can be recorded by observation (mapping, measurement). An extrapolation of the observed behaviour to other states or boundary conditions is usually not possible without consideration of the underlying physical relationships. In order to be able to assess the behaviour of erosion for conditions or boundary conditions other than those given during the measurement (e.g., to derive risk forecasts), a model is required which can describe the interaction of the various individual influences either statistically or on the basis of physical laws.

The development of such models was initially carried out primarily in the USA. This was caused by catastrophic erosion damage, which became increasingly widespread at the beginning of this century, especially in the middle

---

[1] Technical University Freiberg.
    Email jhschmidt@web.de
[2] Geognostics, Berlin.
    Email michael.von.werner@geognostics.de

west of the country. The aim of developing the model was to estimate the long-term soil loss to be expected under current crop conditions and the success of possible countermeasures (in particular with regard to changes in cultivation and tillage methods) on the basis of the most objective principles possible.

The first useful approach to describe water erosion in this respect was the so-called Universal Soil Loss Equation by Wischmeier and Smith (1965). The equation developed on the basis of extensive erosion data describes the average annual erosion as a function of various empirically determined factors with which the influences of climate, soil and agriculture on erosion are mapped.

Purely empirical models such as USLE (Universal Soil Loss Equation), since they are derived from erosion data, allow a forecast of soil loss, however, they are not able to quantify the transport of the excavated material and its deposition elsewhere. Therefore, purely empirical approaches are seldom sufficient for predicting so-called "off-site" losses. Then again, as it is precisely these damages that have gained in importance, newer erosion models make use of predominantly process-oriented, physically based approaches with which the effects of erosion—e.g., substance inputs into the water network—outside the agricultural area can also be calculated. One of the first model systems of this kind was CREAMS (Knisel 1980).

Within the scope of this article, only a few selected approaches can be described from the multitude of existing models for water erosion. The selection is limited to models that are conceptually geared to practical planning applications. Pure research models are not considered.

Empirical Models

The most widely used empirical approach to describe water erosion is the Universal Soil Loss Equation or USLE by Wischmeier and Smith (1965). The equation is based on extensive data from soil erosion measurements carried out between 1930 and 1952 in the Midwest of the USA on standardized erosion measuring plots.

The USLE describes the mean annual soil loss A as the product of the following correlatively determined factors:

$$A = R \cdot K \cdot LS \cdot C \cdot P \tag{1}$$

These factors are characterized by:

- R-factor: the erosive effect of precipitation and surface runoff
- K-factor: the erodibility of the soil under standardized conditions (plot length = 22.6 m, inclination = 9%, fallow land)
- LS-factor: the change in soil loss under standard conditions differing slope length and inclination
- C-factor: the erosion-reducing influence of different crops and processing methods in comparison with fallow land

- *P*-factor: the erosion-reducing effect of protective measures, e.g., contour ploughing, grass strips, etc.

The weighting of the factors in Eq. 1 is determined in the definition of the factors themselves. The most important information contained in USLE is therefore the definition of the factors or the nomograms from which the factor values are derived (Dettling 1989).

The determination equations of the individual factors, initially defined in American units, were transferred into metric units for applications outside the USA. The following explanations refer to the latter.

The **rain factor R** is derived from the kinetic energy of the individual precipitation and its intensity. Information on surface runoff or infiltration rate is not included in the calculation.

The R-factor is determined in several steps. To estimate the kinetic energy E_{kin} (in Joule/m²), the individual precipitation is first divided into any number of (n) subsections (i) with approximately constant intensity (I_i) and the amount of precipitation (N_i) belonging to each subsection is determined. The kinetic energy of the entire rain then results from the combination of these variables, according to Eq. 2 (Dikau 1986):

$$E_{kin} = \sum_{i=1}^{n}(11,89 + 8,73 \cdot \log I_i) \cdot N_i \qquad (2)$$

In a further step, the product of the kinetic energy E_{kin} and the maximum 30-minute intensity I_{30} is calculated for each individual erosive rainfall:

$$EI_{30} = E_{kin} \cdot I_{30} \qquad (3)$$

The sum of the EI_{30} values of all erosion-effective single events of a year results in the R-factor, related to a single year. Due to the high annual fluctuations, the mean of the R-factors of as many individual years as possible should be used to indicate the average erosivity of precipitation.

The **K-factor**—relative to the standard slope (plot length = 22.6 m, slope = 9%, fallow land)—indicates the average soil loss (A) per unit of factor R ($K=A/R$). It is a measure of the resistance of the soil to erosion.

To estimate the K-factor Wischmeier and Smith propose the following regression equation (cf. Schwertmann et al. 1990):

$$K = 2,77 \cdot 10^{-6} \cdot M^{1,14} \cdot (12 - OS) + 0,043 \cdot (A - 2) + 0,0033 \cdot (4 - D) \qquad (4)$$

The equation is based on measurements carried out in the USA (see above) and takes into account the following soil properties which do not vary much over time:

M = content (in %) of grain size class 0,002 ≤ d ≤ 1 mm (corresponds to silt + finest sand)

OS = organic matter content (in %) (*OS=4%* applies for*OS>4%*)

$A =$ aggregate size class
$D =$ permeability class.

Equation 4 is not fully transferable to Central European soil conditions; among other things it only applies to soils with silt and fine sand contents below 70%. Therefore, equation 4 is not applicable for the widespread loess soils with low sand content and clay contents of 10% to 25% which are at risk of erosion in Central Europe (Schramm 1994).

The **LS-factor** describes the change of soil loss with slope length and slope inclination deviating from the standard slope. The influence of the slope shape on the material removal can be taken into account indirectly via weighting factors.

There are two different equations for the determination of the LS-factor, but according to Schwertmann et al. (1990) they yield almost identical results:

$$LS = (l/22)^m \cdot (65{,}41 \cdot \sin^2\alpha + 4,56 \cdot \sin \alpha \cdot 0{,}065) \tag{5}$$

or

$$LS = (l/22)^m \cdot s/9 \cdot \sqrt{(s/9)} \tag{6}$$

where l = erosive slope length in m, α = slope angle in degrees and s = slope angle in %. The exponent m contained in both equations depends on the slope angle (Table 1).

The erosive slope length l is defined as "the length between the place on the slope where on average the surface runoff begins and the place where the deposition of soil material begins on the lower slope" (Schwertmann et al. 1990). Therefore, in Eqs. 5 and 6, the total length of the slope is not included, but only that part which is directly subject to erosion on average of the events.

The **C-factor** takes into account the erosion-reducing effect of soil cover by cultivated plants or harvest residues. The calculation of the C-factor requires that the soil loss occurring with a certain crop and cultivation technique is known as a relative value to the soil loss occurring with fallow land.

The relative soil loss (RSL) is usually a time-dependent variable, as the degree of cover and the soil condition vary during the vegetation period. The RSL values should therefore be differentiated by cultivation period and weighted according to the seasonal distribution of the R-factor. The addition of the weighted RSL values of a year results in the C-factor.

Table 1. Value of m for different slopes.

Slope angle (%)	m
≤ 0.5	0.15
0.6–1.0	0.20
1.1–3.4	0.30
3.5–4.9	0.40
≥ 5.0	0.50

Due to the multitude of possible usage variants and processing methods, the determination of the C-factors is extraordinarily complex. As a rule, they only have regional validity due to the pronounced regional differentiation of land use.

The **P-factor** is used to assess the effect of erosion protection measures. The factor values express the ratio of soil erosion with protective measures to those without protective measures. For certain measures, such as contour tillage or strip usage, the factor values can be taken from tables depending on the slope length and slope inclination or the strip width. For other measures, the P-factors must be determined experimentally.

Purely empirical models such as USLE have the disadvantage that they cannot be transferred to other conditions without restrictions. In any case, the factors included in the equation must first be adapted to regional climate and soil conditions.

Other points of criticism are:

- The definition range of USLE is restricted to the area of the slope that is directly subject to erosion. USLE therefore does not provide any information on the deposition of the removed soil material (location, quantity and particle size distribution) or on its inflow into the water network.
- USLE is not suitable for estimating the erosion of individual erosion events (Foster et al. 1985). It is therefore not possible to derive statements on peak loads or extreme events that are required for risk assessment and as a basis for assessing protective measures.
- The equation cannot be applied to arbitrarily small, homogeneous compartments (Dettling 1989). The influences of the relief, the covering, etc. cannot therefore be adequately considered in the case of differentiated slope sections.
- The effort to determine the USLE factors is disproportionately high, provided that the rules for determining the factors are observed and the necessary local adjustments are actually made.

Physically Based Simulation Models

Due to the inadequacies of purely empirical approaches, more process-oriented, physically based erosion models have been developed. In contrast to empirical models, they are based on physically defined model parameters and are therefore transferable, at least in terms of approach. However, even by means of a physically based model, the real processes can only be reproduced in a very simplified way. This also limits the applicability of these models and results in certain systematic errors.

Irrespective of these limitations, more process-oriented, physically based model approaches promise not only easier transferability but above all improved prediction accuracy and higher spatial and temporal resolution. Particularly in view of the increasing importance of "off-site" damages, they are therefore more suitable for assessing damages or risks associated with erosion and defining requirements for protective measures. However, the practical use of physically based models often fails because the required information about parameters is not completely and reliably available or the handling of the software programs is so complicated that a longer training period is required. Some models, such as OPUS (Smith 1988, 1992), are designed from the beginning as pure research models, so that a broad application in practice is not aimed for anyway.

Table 2 provides an overview of some of the physically based model systems currently available. The majority of the models mentioned here are currently still in the development or test phase. None of the models has been sufficiently validated to date.

The following models will be presented in more detail below: CREAMS (Knisel 1980), WEPP (Lane and Nearing 1989), EROSION 2D (Schmidt 1991) and EUROSEM (Morgan et al. 1992). What these models have in common is that the following process components are differentiated for the mathematical description of erosion:

- the **detachment** of the soil particles from the overflowed surface
- the **transport** of the particles with the surface runoff and
- the **deposition** of particles.

The reference period for the calculation is always individual single events. This considers that the discharge of solids, as already mentioned, is not a continuous process, but is always linked to individual events characterized by different soil and weather conditions.

The mathematical basis of almost all physically based simulation models is the so-called continuity equation. It can be expressed in simplified form as:

$$\frac{\delta q_s}{\delta x} = \gamma(x,t) \qquad (7)$$

Table 2. Overview of some currently available, physically based soil erosion models.

Name	Developer	Spatial reference	Temporal reference
CREAMS	Knisel 1980	Slope (structured)	Individual event
ANSWERS	Beasley and Huggins 1981	Catchment area (structured)	Individual event
OPUS	Smith 1988	Slope (unstructured)	Individual event
WEPP	Lane and Nearing 1989	Slope (structured), Catchment area version in work	Individual event Long term simulation
EUROSEM/ KINEROS	Morgan 1992/ Woolhiser et al. 1990	Slope (structured), Catchment area	Individual event
EROSION 2D/3D	Schmidt 1991	Slope (structured), Catchment area	Individual event (linkable to sequences)

The equation states that the erosion ($\gamma < 0$) or deposition rate ($\gamma > 0$) always corresponds to the change of the solid mass flow (q_s) along the flow path (x). Figure 1 shows the soil erosion and deposition resulting from the course of the solid mass flow to explain this relationship. As the example shows, material is removed as long as the mass flow of solids increases and material is deposited as soon as it decreases. The steeper the curve of the solid mass flow rises or falls, the greater the area-related removal or deposition rate.

Another basic assumption of physically based erosion models is that the discharge by separation cannot be greater than the transport possible at maximum utilization of the surface-parallel flow. If the actual concentration of the particles exceeds the maximum concentration given by the **transport capacity**, e.g., when the flow velocity decreases, the excess proportion of particles suspended in the

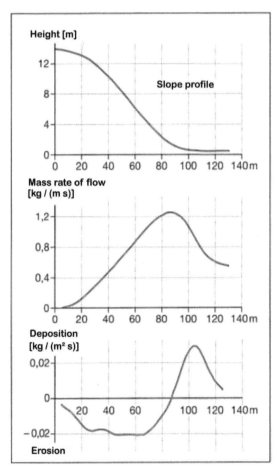

Fig. 1. Mass rate of flow, erosion and deposition on a convex-concave slope profile (simulation: EROSION 2D).

effluent settles again. Erosion or deposition are thus determined either by the properties of the soil (in the form of the erosion resistance to be overcome when the particles are detached) or by the properties of the surface-parallel flow (in the form of the transport capacity).

One of the first physically based erosion models developed on the basis of these conceptual considerations is **CREAMS** (A Field Scale Model for Chemicals, Runoff, and Erosion from Agricultural Management Systems). The model consists of three independent submodels: a hydrological model, an erosion model and a nutrient and pesticide model. The last two submodels each refer to the calculated data of the previous submodel.

The hydrological submodel calculates the runoff at the soil surface on the basis of an infiltration approach developed by Green and Ampt (1911). If the precipitation data are not available in the required time resolution, a simple empirical estimation method (SCS curve number method) is used.

The erosion submodel distinguishes the already mentioned subprocesses: Separation of particles, transport with superficial discharge and deposition of particles.

The **detachment** of the soil particles is calculated with a modified form of USLE separately for inter-rill areas and rills. The **inter-rill erosion** depends on the erosivity of the precipitation EI_{30} and the **rill erosion** depends on the runoff rate q as well as on the slope length x. In the approaches for rill and inter-rill erosion (Eqs. 8 and 9, respectively), the factors K, C and P known from USLE as well as the slope inclination α are also taken into account:

$$D_f = 37983 \cdot m \cdot q_{pk}^{\frac{4}{3}} \cdot \left(\frac{x}{76,2}\right)^{m-1} \cdot (\sin \alpha)^2 \cdot K \cdot C \cdot P \tag{8}$$

$$D_i = 0,21 \cdot EI_{30} \cdot (\sin \alpha + 0,014) \cdot K \cdot C \cdot P \cdot \left(\frac{q_{pk}}{Q}\right) \tag{9}$$

If not explained yet, q_{pk} = peak flow rate, Q = flow volume, m = dimensionless slope length exponent (values in American units!).

The calculation of the **transport capacity** (T_c) of the surface-parallel flow is based on an approach of Yalin (1963). The explanation of this approach in the CREAMS manual (Knisel 1980) is physically only conditionally comprehensible. The following equations were therefore taken directly from the original literature.

Yalin describes the transport capacity T_c (see Eq. 10) as a function of the dimensionless transport coefficient P_s, the density of the particles ρ_s and the liquid ρ_q, the particle diameter D and the shear stress velocity $v*$. According to Eq. 11, the shear stress velocity $v*$ is derived from the acceleration due to gravity g, the slope S and the layer thickness of the runoff δ:

$$T_c = P_s \cdot (\rho_s - \rho_q) \cdot D \cdot v* \tag{10}$$

$$v* = \sqrt{g \cdot S \cdot \delta} \tag{11}$$

The dimensionless transport coefficient P_s contained in equation 10 is defined as:

$$P_s = 0,635 \cdot s \cdot \left[1 - \frac{1}{\sigma} \cdot \log(1+\sigma) \right] \tag{12}$$

The constant 0.635 contained herein is empirically determined; s and σ are further dimensionless parameters. The following apply here:

$$s = \frac{Y}{Y_{crit}} - 1 \tag{13}$$

$$\sigma = 2,45 \cdot \left(\frac{\rho_q}{\rho_s} \right)^{0,4} \cdot \sqrt{Y_{crit}} \cdot s \tag{14}$$

$$Y = \frac{\rho_q \cdot v_*^2}{(\rho_s - \rho_q) \cdot g \cdot D} \tag{15}$$

Y_{crit} is the critical dimensionless shear stress at which erosion begins as a function of the Reynolds number of particles ($X = (D \cdot v^*)/v$). Y_{crit} must be determined experimentally or estimated from existing data (e.g., the Shields diagram).

Equation 10 applies in the form given here only to equal grain sediments. In CREAMS, the different particle sizes of a grain mixture are taken into account by assigning a certain proportion of the transport capacity to each particle class (a total of 5) depending on the size of the particles and their specific weight. If the proportionate transport capacity in one of the particle classes is greater than the quantity of sediment actually transported in this class and if there is a surplus of sediment in another class at the same time, the excess transport capacity shall be allocated to the first class of the second class. The share of the second particle class in the available transport capacity can thus be increased relative to the share of the other classes (cf. Astalosch 1990).

Particle detachment and transport are calculated segment by segment from the highest point of the slope to the base of the slope.

For discharge from a segment, the following are decisive:

- the input into the segment with the inflow from above
- the amount of sediment detached within the segment per unit of time and unit of area (depending on the erosivity of the precipitation, the runoff rate and the specific erodibility of the soil)
- the transport capacity

In this case, only as much soil can be removed in a segment as can also be removed with transport capacity. If the input into the segment is already greater than the transport capacity, the portion exceeding the transport capacity is deposited within the segment. The area-related deposition rate D_u is calculated according to:

$$D_u = \varepsilon \cdot \frac{v_s}{q} \cdot (T_c - G) \tag{16}$$

Here ε is a dimensionless, empirically derived coefficient, v_s the sinking velocity of the particles, q the flow rate, T_c the transport capacity and G the input into the segment of upper current.

Due to the consideration of the sinking velocity in Eq. 16, CREAMS allows a fractional differentiation of the sedimented particles in the case of deposition. This distinction is of great importance for assessing possible "off-site" damages, as the fine soil particles are generally much more contaminated than the coarser ones. As soon as deposition occurs in one of the segments, CREAMS recalculates the grain size distribution of the transported sediment. This takes into account that when the transport capacity is exceeded, the coarser particles preferably settle, while the finer particles accumulate in the sediment. However, CREAMS cannot describe the selective detachment of soil particles. An enrichment of the finer fractions can therefore only occur after particles have already sedimented.

The advantages of the CREAMS model lie primarily in the possibility of differentiating erosion events in terms of time and location and in the fraction-dependent consideration of deposition. In addition, CREAMS provides information on the discharge of particle-bound pollutants and nutrients. However, restrictions result from the use of elements of USLE (see above). Another critical factor is the use of the Yalin equation to describe the transport capacity. The experimental data that led to this equation were obtained under fluidized bed conditions of equal grain sediments. It is doubtful whether the Yalin equation can be transferred to the specific conditions of mass transport of inhomogeneous source substrates by essentially sheet-like flows (Schramm 1994). As can be shown in the experiment, the hydraulic properties of runoffs at shallow flow depths are influenced to a much greater extent by precipitation than is the case at greater water depths (Yoon and Wenzel 1971). This effect is not included in the Yalin equation (Guy and Dickinson 1990).

A similar approach to the CREAMS model to describe erosion is used in WEPP (Water Erosion Prediction Project). WEPP (Foster and Lane 1987, Lane and Nearing 1989, Flanagan 1990) is currently still in the test or development phase. Unlike CREAMS, this model no longer uses USLE elements.

The WEPP model allows both an estimation of erosion related to individual single events and a prognosis over a longer period (several years). The behaviour of the relevant parameters for erosion (plant cover, erosion resistance, soil moisture, etc.), is continuously mapped here, including the time periods between the erosion-effective individual events. The initial and boundary conditions must therefore only be entered once at the beginning of the simulation. The model then calculates the seasonal changes in the relevant values itself. The climate parameters precipitation, temperature and radiation can also be generated automatically via a specific component of the model.

The hydrological submodel of WEPP describes the relationship between precipitation and runoff (just like CREAMS) on the basis of a modified Green and Ampt approach. The excess precipitation is divided into sheet and rill flow, so that, similar to CREAMS, a distinction can be made between rill and inter-rill erosion.

The detachment of the soil particles by rill flow is expressed by:

$$D_c = k_r \cdot (\tau_f - \tau_c) \tag{17}$$

In this context: D_c the detachment capacity of the flow, k_r the specific erodibility of the rill bottom, τ_f the shear stress exerted on the soil particles by the rill flow and τ_c the critical shear resistance of the soil.

The shear stress τ_f is derived according to Eq. 18 from the specific gravity of water γ ($\gamma = \rho\, g$), the mean gradient S and the hydraulic radius R (based on a rectangular rill cross-section):

$$\tau_f = \gamma \cdot S \cdot R \cdot (f_s/f_t) \tag{18}$$

In the case that the soil is covered by plants or similar, the quotient f_s/f_t takes into account that part of the shear stress is consumed by plant parts, etc., in the flow cross-section.

Taking into account the efficiency of the flow, given by the ratio of the sediment quantity G (see below) to the transport capacity T_c, the net discharge D_f in the rills results according to:

$$D_f = D_c \cdot \left[1 - \frac{G}{T_c} \right] \tag{19}$$

Regardless of the detachment capacity of the flow (D_c), the net discharge (D_f) is zero as soon as the amount of sediment (G) carried along by the flow equals the transport capacity (T_c). For the determination of the transport capacity T_c a modified version of the YALIN equation is used (as with CREAMS).

The erosion contribution of the inter-rill areas is derived according to Eq. 20 from: the specific erodibility K_i of the inter-rill areas, the effective precipitation intensity I_e, the coefficients C_e and G_e (to indicate the influence of plant or soil cover), the groove spacing R_s and the rill width W:

$$D_i = K_i \cdot I_e^2 \cdot C_e \cdot G_e \cdot \left(\frac{R_s}{W} \right) \tag{20}$$

The total discharge is then calculated by adding the contributions from rill (D_f) and inter-rill areas (D_i).

Deposition occurs when the sediment quantity $G = (D_f + D_i)x$ (x = slope length) exceeds the transport capacity T_c of the rill flow. This case is described by:

$$D_u = \beta \cdot \frac{v_s}{q} \cdot (T_c - G) \tag{21}$$

Equation 21 fully corresponds to the approach chosen in CREAMS. It means that vs is the falling velocity of the sediment particles suspended in the flow and q the discharge related to the slope width. The other variables have already been explained.

WEPP is capable of continuously modeling the behavior of numerous parameters relevant for erosion over a longer period of time, beyond the contexts described here in summary. This requires, among other things, a much more comprehensive description of the soil water balance. For these reasons, the model includes further model approaches—for example to describe plant growth, evaporation, drainage and snow melt. However, this also increases the need for data that must first be entered into the model. This circumstance may severely restrict the application of the model in practice. This disadvantage, which—although not to the same extent—also applies to other physically based approaches, is counterbalanced by the advantage that the model should be transferable to other conditions without extensive adjustments.

The model **EROSION 3D** was developed with the intention to create an easy-to-use tool for erosion prediction in soil and water conservation planning and assessment (Schmidt 1992, Von Werner 1995). The model, which is predominantly based on physical principles, simulates the detachment of soil, the transport deposition of detached soil particles by overland flow, incl. the grain size distribution of the transported sediment and the sediment delivery into downstream water courses caused by single events (Schmidt 1992).

The theoretical base of the model was initially developed by J. Schmidt (1991, 1996) and later extended by M. Von Werner (1995). The model calculations are executed for small and homogenous spatial raster elements and temporal steps, allowing the model to perform simulations with a high spatial and temporal resolution.

The model is structured along two main sub-models referring to infiltration and runoff, soil detachment and transport (Fig. 2). Based on a specific spatiotemporal element the simulation starts with the calculation of excess rainfall (infiltration sub-model) followed by the analysis of flow distribution (kinematic flow routing). Detachment and transport resp. deposition of soil particles are then calculated on the basis of the momentum fluxes exerted by falling raindrops resp. surface runoff.

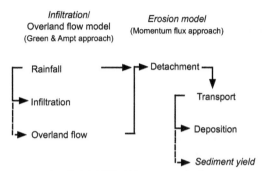

Fig. 2. EROSION 3D conceptual scheme.

Infiltration sub-model

The infiltration submodel of EROSION 3D is based on the approach of Green and Ampt (1911) which includes a simplification of the infiltration process by assuming that rain water penetrates the soil in a piston-like flow and saturates the available pore space completely. For the mathematical description, the infiltration process is divided into a gravitational component i_1 and a dynamic matric component i_2 (Weigert and Schmidt 2005). The gravitational component i_1 is a function of the gravitational potential Ψ_g.

$$i_1 = k \cdot \frac{\Delta \Psi_g}{x_{f1}} = k \cdot g \qquad (22)$$

where i_1 = infiltration rate of the gravitational component [kg/(m² s)], k = hydraulic conductivity of the transport zone [(kg s)/m³], Ψ_g = gravitational potential [(N m)/kg], x_{f1} = depth of the wetting front of the gravitational component [m], g = gravity constant [m/s²]

The matric component i_2 is a function of the matric potential Ψ_m.

$$i_2 = k \cdot \frac{\Delta \Psi_m}{x_{f2}(t)} \qquad (23)$$

where i_2 = infiltration rate of the matric component [kg/(m² s)], k = hydraulic conductivity of the transport zone [(kg s)/m³], Ψ_m = matric potential [(N m)/kg], $x_{f2}(t)$ = depth of the wetting front of the matric component [m] at time t.

Assuming a continually advancing wetting front that moves downward in the soil, the volume of water infiltrating the soil during a particular time interval is a product of penetration velocity (dx_f/dt) multiplied by the difference of initial and saturated soil water content. Hence, the gravitational component i_1 can be calculated as follows:

$$i_1 = k \cdot g = \rho_f \cdot \Delta\Theta \cdot \frac{dx_{f1}}{dt} \tag{24}$$

with

$$\Delta\Theta = \Theta_s - \Theta_0$$

where i_1 = infiltration rate of the gravitational component [kg/(m² s)], k = hydraulic conductivity of the transport zone [(kg s)/m³], g = gravity [m/s²], ρ_f = fluid density [kg/m³], x_{f1} = depth of the wetting front of the gravitational component [m] at time t, t = time [s], Θ_s = saturated water content [m³/m³], Θ_0 = initial water content [m³/m³].

Similarly, the matric component i_2 is given by

$$i_2 = k \cdot \frac{\Delta\Psi_m}{x_{f2}(t)} = \rho_f \cdot \Delta\Theta \cdot \frac{dx_{f2}}{dt} \tag{25}$$

with

$$\Delta\Psi_m = \Psi_{m0} - \Psi_{ms}$$

where i_2 = infiltration rate of the matric component [kg/(m² s)], k = hydraulic conductivity of the transport zone [(kg s)/m³], ρ_f = fluid density [kg/m³], $x_{f2}(t)$ = depth of the wetting front of the matric component [m] at time t, t = time [s], Θ_s = saturated water content [m³/m³], Θ_0 = initial water content [m³/m³], Ψ_{m0} = matric potential related to the initial water content Θ_0 [N m/kg], Ψ_{ms} = matric potential related to the water content of the transport zone Θ_s [N m/kg].

Under the assumption of nearly saturated conditions within the transport zone the simplification $\Psi_{ms} \approx 0$ can be made, so that $\Delta\Psi_m \approx \Psi_{m0}$ and $k \approx k_s$.

By rearranging and integrating Eqs. (25) and (26), the depth of the wetting front x_{f1} at time t for the gravitational component i_1 is obtained by:

$$x_{f1} = \frac{k_s \cdot g \cdot t}{\rho_f \cdot (\Theta_s - \Theta_0)} \tag{26}$$

where x_{f1} = depth of the wetting front of the gravitational component [m], k_s = saturated hydraulic conductivity [(kg s)/m³], g = gravity constant [m/s²], t = time [s], ρ_f = fluid density [kg/m³], Θ_s = saturated water content [m³/m³], Θ_0 = initial water content [m³/m³].

And for the matric component i_2 by

$$x_{f2} = \sqrt{\frac{2k_s \cdot \Psi_{m0} \cdot t}{\rho_f \cdot (\Theta_s - \Theta_0)}} \tag{27}$$

where x_{f2} = depth of the wetting front matric component [m], k_s = saturated hydraulic conductivity [(kg s)/m³], Ψ_{m0} = matric potential related to the initial water content Θ_0 [N m/kg], t = time [s], ρ_f = fluid density [kg/m^3], Θ_s = saturated water content [m^3/m^3], Θ_0 = initial water content [m^3/m^3].

Hence x_{f1} (Eq. 26) can be inserted into equation 22 and x_{f2} (Eq. 27) into Eq. 23:

$$i_1 = \frac{\Delta\Psi_g}{\dfrac{g \cdot t}{\rho_f \cdot (\Theta_s - \Theta_0)}} = k_s \cdot g \tag{28}$$

$$i_2 = k_s \cdot \frac{\Psi_{m_0}}{\sqrt{\dfrac{2k_s \cdot \Psi_{m_0} \cdot t}{\rho_f \cdot (\Theta_s - \Theta_0)}}} \tag{29}$$

Now the infiltration rate can be calculated as the sum of the gravitational i_1 and the matric component i_2:

$$i = i_1 + i_2 = k_s \cdot g + k_s \cdot \frac{\Psi_{m_0}}{\sqrt{\dfrac{2k_s \cdot \Psi_{m_0} \cdot t}{\rho_f \cdot (\Theta_s - \Theta_0)}}} \tag{30}$$

where i = infiltration rate [kg/(m^2 s)], i_1 = infiltration rate of the gravitational component [kg/(m^2 s)], i_2 = infiltration rate of the matric component [kg/(m^2 s)], k_s = saturated hydraulic conductivity [(kg s)/m^3], g = gravity constant [m/s^2].

The independent variables of this equation can either be directly estimated from field measurements (i.e., the initial water content θ_0), or be derived from basic soil parameters by applying the following pedotransfer functions:

$$k_s = 4 \cdot 10^{-3} (1{,}3 \cdot 10^{-3} / \rho_b)^{1{,}3 \cdot b} \cdot \exp(0{,}069 \cdot T - 0{,}037 \cdot U) \tag{31}$$

with

$$b = (10^{-3} \cdot D)^{-0{,}5} + 0{,}2 \cdot \delta_p \quad \text{(Campbell 1985)}$$

where k_s = saturated hydraulic conductivity [(kg s)/m^3], ρ_b = bulk density [kg/m^3], T = clay content [kg/kg], U = silt content [kg/kg], b = parameter [–], D = mean diameter of soil particles [m], δ_p = standard derivation of the mean diameter of soil particles [–].

$$\Psi_{m_0} = \frac{\left[\left(\dfrac{\Theta_s - \Theta_r}{\Theta_0 - \Theta_r} \right) \cdot \dfrac{1}{\alpha^n} \right]^{\frac{1}{n}}}{100 \cdot \rho_b} \quad \text{(Van Genuchten 1980)} \tag{32}$$

where Ψ_{m_0} = matric potential related to the initial water content θ_0 [N m/kg], ρ_b = bulk density [kg/m^3], θ_0 = initial water content [m^3/m^3], θ_r = residual water content [m^3/m^3], θ_s = saturated water content [m^3/m^3], θ, n = parameters [–].

Since the theoretical concept of infiltration presupposes a rigid soil matrix, time-variable structural processes such as soil compaction, slaking and crusting of macropores due to shrinking and biological activities have to be considered by an empirical factor, called skinfactor. This factor allows calibrating the saturated hydraulic conductivity k_s according to equation (31) on the basis of experimentally measured data (Michael 2000). Values of skinfactor < 1 reduce infiltration rate, in order to take into account the effects of soil slaking and crusting as well as anthropogenic compaction. Values of skinfactor > 1 cause a positive correction of infiltration rate, e.g., for the consideration of an increased infiltration in macropores due to soil shrinking, biological activity or tillage impact. If skinfactor = 1 infiltration rate is obviously not affected by either slaking and sealing or macropores.

Heavy rainstorms are mostly short-term phenomena. Consequently, the wetting front only penetrates a few decimeters deep into the soil, so that only the properties of the uppermost soil layer have to be considered in the simulation. However, in the case of long-term rainstorms, the modeling approach should also consider deeper soil layers in their influence on infiltration. To do so, EROSION 3D provides a parameterization interface and appropriate modeling routines. In order to calculate infiltration, the smallest hydraulic conductivity within the transport zone is used, while the matrix potential driving the infiltration flow is always taken directly in front of the wetting front.

In addition EROSION 3D has further model approaches that take into account the influence of trapped air and water repellency. These factors can significantly reduce infiltration at very low initial soil moisture levels. Frozen soil has a similar impact on infiltration, which can also be considered in the EROSION 3D model (Weigert and Schmidt 2005).

Detachment and transport sub-models

The theoretical concept of the detachment sub-model is based on the assumption that the erosive impact of overland flow and rainfall droplets is proportional to momentum fluxes exerted by the flow and the falling droplets respectively (Schmidt 1991 and 1992), defined in general form by:

$$\varphi = \frac{m \cdot v}{t}; \left[\frac{kg \cdot m}{s \cdot s} = \frac{kg \cdot m}{s^2} = N \right] \tag{33}$$

where m/t = mass rate of surface runoff respectively rainfall, v = runoff velocity respectively fall velocity of droplets. As Fig. 3 shows, momentum fluxes shall be regarded as vectors.

Assuming macroscopic perspective erosion occurs if the sum of all mobilizing forces acting on the soil particles (given by the momentum fluxes of surface runoff φ_q and raindrops φ_r) is greater than the sum of those forces that prevent particles from being detached and transported. In all other cases, no particles are eroded

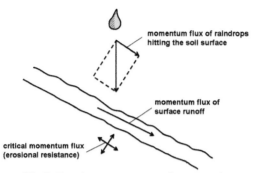

Fig. 3. Detachment—momentum flux approach.

from the soil surface. Following this concept the erosional effects of raindrops and overland flow can be related to the soil's resistance to erosion (given by the critical momentum flux φ_{crit}) to give a dimensionless coefficient E (Eq. 34):

$$E = \frac{\varphi_q + \varphi_r \cdot \sin \alpha}{\varphi_{crit}} \tag{34}$$

where φ_q = momentum flux exerted by surface runoff [N], φ_r = momentum flux exerted by raindrops [N] and $\varphi_{crit.}$= critical momentum flux (erosional resistance) [N].

Erosion occurs if $E > 1$ whereas $E \leq 1$ characterizes the erosion-free state of flow.

For quantitative results, the coefficient E is correlated with experimental data. Fifty experiments under simulated rainfall have been performed in a test flume filled with silty soil (Schmidt 1988). The data can be fitted by the following regression equation:

$$q_s = (1,75E - 1,75) \cdot 10^{-4} \tag{35}$$

where q_s = sediment discharge of detached particles. Figure 4 shows the regression curve and the experimental data on which the curve is based. Because of the theoretical postulate that sediment cannot be eroded when $E \leq 1$ the regression curve muss intersect the x-axis at $E = 1$.

The momentum flux exerted by raindrops is defined as (15):

$$\varphi_r = r_a \cdot \rho_r \cdot v_r \cdot A \cdot \sin \alpha \cdot (1 - C_L) \tag{36}$$

where φ_r = momentum flux exerted by raindrops [N], $r_a = r \cos\alpha$ = effective rainfall intensity [m/s] related to the slope surface, ρ_r = fluid density of rainwater [kg/m³], v_r = mean fall velocity of raindrops, A = area of the slope segment [m²], α=slope angle and C_L= ground cover. The effective rainfall intensity r_a is introduced to Eq. 24 because by default rainfall intensity data refer to a horizontal plane. To calculate runoff generation and soil detachment these intensity data have to be transferred to inclined surfaces.

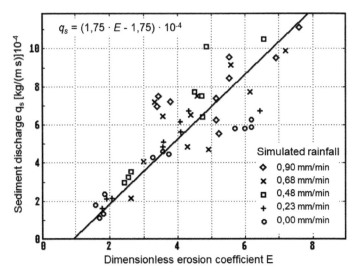

Fig. 4. Measured sediment discharge q_s vs. erosion coefficient E.

The fall velocity of raindrops contained in Eq. 36 is very difficult to measure under field conditions. Available data show that the size and hence the velocity of the droplets increase with rainfall intensity (Laws 1941, Laws and Parsons 1943). By making use of these data, we obtain the following empirical equation (Eq. 37), which provides a simple method of estimating the mean fall velocity of raindrops on the basis of rainfall intensity data.

$$v_r = 4{,}5 \cdot r^{0,12} \tag{37}$$

where v_r = mean fall velocity of raindrops [m/s], r = rainfall intensity [mm]
    In analogy to Eq. 36 the momentum flux exerted by overland flow is determined by:

$$\varphi_q = q \cdot \rho_q \cdot v_q \cdot \Delta y \tag{38}$$

where φ_q = momentum flux exerted by flow [N], q = volume rate of flow [m³/(m s)], ρ_q = fluid density [kg/m³], Δy = the width of the slope segment [m] and v_q = mean flow velocity [m/s] according to the Manning equation:

$$v_q = \frac{1}{n} \cdot \delta^{2/3} \cdot S^{1/2} \tag{39}$$

where n = coefficient of surface roughness [s m$^{-(1/3)}$], δ = thickness of runoff water film [m], S = slope.
    In order to transport detached particles the uplift by vertical (turbulent) flow components of surface runoff must counteract the gravitational settling

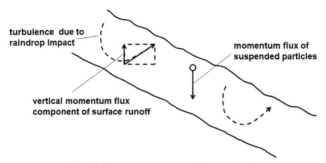

Fig. 5. Transport-momentum flux approach.

of the suspended particles (Fig. 5). Since surface runoff in this case is usually developed as a thin water film in the range of millimeters up to some centimeters in depth, flow turbulence is predominantly a result of raindrop impact and not due to friction effects within the water film. Raindrop impact results in an irregular motion of surface runoff, which is essential for the lift up of eroded particles and particle transport in suspension. Without raindrop impact and consequently without turbulence only bedload transport occurs which is far less effective than sediment transport in suspension.

To transfer the concept of particle transportation consistently to the momentum flux approach, the vertical momentum flux component of the (turbulent) flow on the one side and the critical momentum flux of particles (which is according to Eqs. 40 and 41 a function of particle size, fluid density and fluid viscosity) on the other side have to be taken under consideration.

$$v_p = \frac{1}{18} \cdot \frac{g \cdot D^2 (\rho_p - \rho_q)}{\eta} \tag{40}$$

where v_p = settling velocity of suspended particles [m/s], D = particle diameter [m], ρ_p = particle density [kg/m³], ρ_q = fluid density [kg/m³], g = acceleration of gravity [m/s²], η = fluid viscosity [kg/(m·s)].

$$\varphi_{p,\,crit} = c \cdot \rho_p \cdot A \cdot v_p^{\,2} \tag{41}$$

where $\varphi_{p,crit}$ = critical momentum flux of suspended particles [(kg m)/(s² m²)], c = concentration of particles [m³/m³], ρ_p = particle density [kg/m³], A = area of slope segment [m²], v_p = settling velocity of soil particles [m/s]

Hence the prior condition for particle transport is given by Eq. 42:

$$\varphi_{q,vert.} \geq \varphi_{p,crit.} \tag{42}$$

where $\varphi_{q,vert.}$ = vertical momentum flux component of surface runoff [N]. $\varphi_{p,crit.}$ = momentum flux of suspended particles [N].

Transport capacity has been reached, when the vertical momentum flux component of the flow equals the critical momentum flux of the suspended particles.

The concentration of particles at transport capacity can be expressed as:

$$c_{max} = \frac{1}{\kappa} \frac{\varphi_q + \varphi_r}{\rho_p \cdot A \cdot v_p^2} \tag{43}$$

where c_{max} = concentration of particles at transport capacity [m³/m³], κ ($\approx$1000) = empirical factor, φ_q = momentum flux exerted by flow [N], φ_r = momentum flux exerted by raindrops [N], ρ_p = particle density [kg/m³], A = area of slope segment [m²], v_p = settling velocity of soil particles [m/s].

Transport capacity is then determined according to:

$$q_{s,max} = c_{max} \cdot \rho_p \cdot q \tag{44}$$

where $q_{s,max}$ = sediment discharge at transport capacity [kg/(m s)], c_{max} = concentration of particles at transport capacity [m³/m³], ρ_p = particle density [kg/m³], q = volume rate of flow [m³/(m s)].

According to Eq. 44, it is possible to calculate the transport capacity for any particle size class separately. The transport capacities derived in this way specify the maximum mass rate of particles that can be transported within this size class under the given flow conditions (assuming that transport is not limited by detachment). In order to determine the actual mass rate and the particle size distribution of the transported sediment, the following assumptions are made: (a) The particle size distribution of the detached sediment is the same as in the original soil. (b) Detachment occurs only if there is excess transport capacity.

This means that the particle size distribution of the transported sediment corresponds to that of the initial soil, as long as the mass rate of the detached particles does not exceed the transport capacity in any of the particle classes considered. If that is not the case, the mass rate of the particles and hence the size distribution of the transported sediment is controlled by transport capacity.

In order to calculate the rate of erosion or deposition for each of the individual slope segments the following simple equation is used:

$$\gamma = \left(\frac{q_{s,in} - q_{s,out}}{\Delta x} \right) \tag{45}$$

where γ = is the rate of erosion ($\gamma < 0$) or deposition ($\gamma > 0$) per unit area, $q_{s,in}$ = the sediment discharge into the segment from the segment above, $q_{s,out}$ = the sediment discharge out of the segment and Δx = the length of the slope segment.

Overland flow and sediment routing

Since the model operates on a grid-cell basis EROSION 3D allows generating drainage paths by which water and sediment can be routed from the top to the bottom of the respective catchment. According to this, the following procedure is used (Von Werner 1995, Seidel and Schmidt 2009): In the first step slope angle and aspect are determined for each grid element. Then all neighboring elements are examined to select those which have lower elevations. Flow distributions are calculated either by directing all runoff to the lowest neighboring element (D8 algorithm) or by distributing runoff to all neighboring elements with lower elevation in proportion to the difference in altitude (FD8 algorithm). For sheet flow conditions the FD8 algorithm yields much better results as it shows a more natural flow distribution compared to the D8 algorithm. However, in case of channel flow the D8 algorithm is preferable because runoff is always directed totally to only one downstream element. In order to differentiate sheet and channel flow runoff, elements are stored in two grid layers. One layer contains the sheet flow whereas the other one holds the channel flow data. The user can classify channel elements manually or automatically by setting up the minimum drainage area upstream that is necessary to define a channel (= critical source area). By specifying a pour point EROSION 3D automatically determines the watershed and the drainage network based on either the FD8 (sheet flow) or the D8 (channel flow) algorithm.

As shown in Fig. 6 there is a water film establishing at the soil surface when runoff occurs according to Eq. 46:

$$\delta = \left(\frac{q \cdot n}{S^{1/2}}\right)^{3/5} \tag{46}$$

where q = volume rate of flow [m³/(m s)], n = coefficient of surface roughness [s m$^{-(1/3)}$], δ = thickness of runoff water film [m], S = slope.

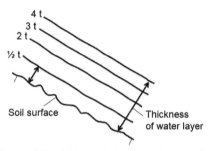

Fig. 6. Change of film thickness during formation of surface runoff.

This water film can be considered as a dynamic storage, which results from the thickness of the film and the area of the overflowed slope segment. Since any change in flow rate results in an equivalent change in film thickness, the film storage fills up as the flow rate increases and empties as it decreases. Major changes in the flow rate occur predominantly at the beginning and at the end of a rainstorm, causing a shift in the runoff hydrograph. Further, the emptying process might maintain runoff beyond the end of the rainfall event (so-called afterflow). In applying this concept, EROSION 3D is capable to simulate a runoff hydrograph at any point of a catchment as shown in Fig. 7.

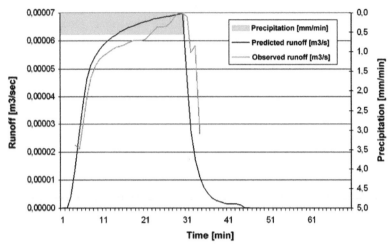

Fig. 7. Comparison of observed and predicted runoff hydrograph.

Input and output parameters

The model's input and output parameters are summarized in Table 3. The input parameters can be assigned to three main groups: relief parameters, surface respective soil parameters and precipitation parameters. The input maps are shown in Fig. 8.

The values of all input parameters are assumed to be spatially uniform below the scale of grid resolution which may vary between 0.1 m and 30 m. The effects of different types of land use and agricultural management practices on erosion are accounted for by varying the values for erosional resistance, hydraulic roughness and percentage soil cover. All spatially distributed data—relief, surface, soil and land use data—are imported from a Geographical Information System data base (e.g., ArcGIS), but direct data access is still possible within the model without using any external software.

Most of the input variables are commonly accessible except the following model-specific parameters: skin factor, surface roughness and resistance to erosion. In principle, these parameters have to be determined by simulated rainfall experiments. However, there is a comprehensive data base obtainable which can be used in combination with the additional software tool DPROC (Schindewolf and Schmidt 2010) to identify EROSION 3D model inputs almost automatically from generally available European Community datasets on soil, land use and soil management.

The model produces raster-based quantitative estimates of excess rainfall, runoff, soil loss/deposition and the sediment delivery (mass rate) into the surface water system. The model outputs, e.g., the predicted spatial distribution of erosion and deposition, can be displayed and plotted as colored maps (Fig. 9) or stored in ASCII-files for further processing. The model results have been extensively

Table 3. EROSION 3D input and output parameter.

Input parameters	Output parameters
Relief parameters: - Digital elevation data *Surface and soil parameters*: - Texture - Bulk density - Organic matter content - Initial soil moisture - Hydraulic roughness (Manning's n) - Resistance to erosion (crit. momentum flux) - Canopy cover - Skin factor *Precipitation parameters*: - Rainfall intensity - Duration and date	*Related to the cross-section of a selected grid element (e.g., the element at the catchment outlet)*: - Runoff (volume rate) - Sediment discharge (mass rate) - Grain size distribution of the transported sediment *Related to the catchment of a selected grid element*: - Erosion/Deposition - Net Erosion

Digital elevation model

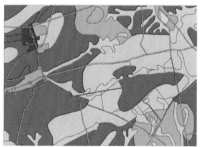

Soil texture map

Landuse map

Fig. 8. EROSION 3D model inputs.

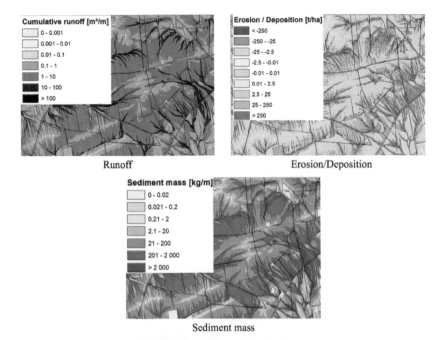

Runoff   Erosion/Deposition

Sediment mass

Fig. 9. EROSION 3D model output maps.

validated. For the application of EROSION 3D there are extensive experiences, especially in Europe, but there are also some for Africa (e.g., Gebreselassie 2012) and Asia.

Application of the EROSION 3D Model

EROSION 3D model has been applied in different parts of the world to test its applicability under different hydrological conditions. The EROSION-3D model was applied by Schmidt et al. (1999) in CATSOP catchment of Netherlands from a period of 1987–1993. The computation results in a raster map showed the spatial pattern of predicted erosion and deposition within the catchment. The model results were compared with observed sediment data measured at the catchment outlet. The comparison showed that simulated soil loss was generally too high which can be due to weak input data. It implied that the input data required for the simulation of soil erosion should be of good quality.

Schob et al. (2006) used the EROSION-3D model in the Saxonian loess belt in Germany for erosion prediction. The EROSION-3D model performed well while predicting the sediment yield from the study watersheds and acted as decision support system to locate the main areas of soil loss and deposition.

Mengistu et al. (2012) applied the two physically based models, EROSION 3D and WEPP for predicting the watershed scale sediment and runoff. Watershed scale modeling results showed that sediment yield and runoff vary by slope, land use and soil type. The spatial sediment budget also showed the variability in the erosion and deposition of sediment in the basin. Average simulated erosion in cultivated land was about 120 t/ha/yr and the lowest simulated erosion rate was on bush lands and grasslands, which indicate that change in land use has significant impact on soil erosion in the Mara River basin.

Starkloff (2012) used two hydrological models LISEM and EROSION-3D for the erosion prediction. EROSION-3D consolidated the results of LISEM calibration and despite the different approaches taken to simulate the surface discharge the results are not significantly different. Due to relatively less input data requirement for the EROSION-3D model, the operational hours of the EROSION-3D are lesser than the LISEM model. Also, the spatial distribution of erosion deposition predicted by the EROSION-3D model appeared to be more accurate than the LISEM model. However, during the calibration of the EROSION-3D model in the sub catchment it was observed that process of finding the correct grid size and time resolution for small catchments is not easy and requires experience.

Kenderessy (2012) used EROSION-3D in his study in Bratislava to locate the main areas of soil loss and to simulate the erosion rates before and after the application of soil protection measures. The results showed that applied measures can effectively reduce soil loss rates and they also reinforce that simulation models such as EROSION-3D are able to provide the information necessary for appropriate localization and extent of site-specific measures.

Schindewolf et al. (2015) applied the EROSION-3D model for prediction of soil erosion in a reservoir in Germany. The EROSION-3D model was successfully applied to simulate reservoir siltation in a meso-scaled German loess catchment. The EROSION-3D-based soil loss prediction maps helped to identify the most erosion-sensitive areas within the catchment, as well as the points of sediment transfer into surface water bodies. It was concluded that the soil conservation measures should be implemented within the catchment to avoid excessive siltation.

Honek et al. (2017) applied the EROSION-3D model to calculate the potential soil water erosion in a small catchment in the Myjava Hill Land, and the model was successfully calibrated for Slovak soil conditions. The results showed the strong interaction between soil condition and potential soil water erosion, corresponding to relief and precipitation.

Lenz et al. (2018) conducted a study to evaluate some important erosion parameters (surface roughness, skin factor and resistance to erosion) required to apply the EROSION-3D model. In this study, four rainfall experiments, each including dry and wet run, were conducted on different land use conditions on a research farm of the Regional Research Station Ballowal Saunkhri. The erosion parameters were evaluated by using two methods and the evaluated values were in close range to experimentally determined values for resistance to erosion and surface roughness.

References

Astalosch, R. 1990. Beregnungsversuche zur Parameterbestimmung zeitlich differenzierter Erosionsmodelle. Unveröffentlichte Vertiefungsarbeit am Institut für Wasserbau und Kulturtechnik der Universität Karlsruhe.

Beasley, D.B. and L.F. Huggins. 1981. ANSWERS—user manual. EPA-Report, 905/9-82-001, Chicago.

Campbell, G.S. 1985. Soil physics with BASIC. Transport models for soil-plant systems. Elsevier Science B.V., Amsterdam, The Netherlands.

Dettling, W. 1989. Die Genauigkeit geoökologischer Feldmethoden und die statistischen Fehler qualitativer Modelle. Physiogeographica - Basler Beiträge zur Physiogeographie 11, Basel.

Dikau, R. 1986. Experimentelle Untersuchungen zu Oberflächenabfluß und Bodenabtrag von Meßparzellen und landwirtschaftlichen Nutzflächen. Heidelberger Geographische Arbeiten 81, Heidelberg.

Flanagan, D.C. 1990. Water Erosion Prediction Project—hillslope profile model documentation corrections and additions. NSERL Report 4 (USDA–ARS National Soil Erosion Laboratory), West Lafayette, Indiana (USA).

Foster, G.R., J.M. Laflen and C.V. Alonso. 1985. A replacement for the Universal Soil Loss Equation (USLE). *In*: DeCOURSEY (ed.): Proceedings of the Natural Resources Modeling Symposium, USDA, ARS-30, S. 468–472, Pingree Park, CO.

Foster, G.R. and L.J. Lane. 1987. User requirements. USDA-Water erosion prediction project (WEPP). National Soil Erosion Research Laboratory Report, 1, West Lafayette, Indiana (USA).

Gebreselassie, B.Y. 2012. Possibilities to Reduce Suspended Sediment Loads into Reservoirs: A Case Study for a Single Reservoir in the Catchment of Kulfo River, Ethiopia. PhD Thesis, Technical University, Dresden.

Green, W.H. and G.A. Ampt. 1911. Studies on soil physics. I: The flow of air and water through soils. J. Agric. Sci. 4: 1–24.

Guy, B.T. and W.T. Dickinson. 1990. Inception of sediment transport in shallow overland flow. Catena (Suppl.) 17: 91–109.

Honek, D., Z. Nemetova and T. Latkova. 2017. Application of physically-based EROSION-3D model in small catchment. Proc 17th International Multidisciplinary Scientific Geo Conference SGEM at Vienna, Austria.

Kenderessy, P. 2012. Soil loss assessment in an agricultural landscape and its utilization in landscape planning. Ekol. (Bratislava) 31: 309–321. DOI: 10.4149/ekol-2012-03-309.

Knisel, W.G. 1980. CREAMS-A field scale model for chemicals, runoff, and erosion from agricultural management systems. Conservation Research Report 26 (USDA), Washington.

Lane, L.J. and M.A. Nearing. 1989: USDA—Water Erosion Prediction Project: hillslope profile model documentation. NSERL Report 2 (USDA-ARS National Soil Erosion Laboratory), West Lafayette, Indiana (USA).

Laws, J.D. 1941. Measurements of the fall-velocity of water-drops and raindrops. Trans. Am. Geophys. Union. 21: 709–721.

Laws, J.D. and D.A. Parsons. 1943. The relation of raindrop size to intensity. Trans. Am. Geophys. Union. 24: 52–460.

Lenz, J., A. Yousuf, M. Schindewolf, M.V. Werner, K. Hartsch, M.J. Singh and J. Schmidt. 2018. Parameterization for EROSION-3D model under simulated rainfall conditions in lower shivaliks of India. Geosciences 8: 1–18.

Mengistu, D., M. Assefa and M.V. Michael. 2012. Watershed scale application of WEPP and EROSION-3D models for assessment of potential sediment source areas and runoff flux in the Mara River Basin, Kenya. Catena 95: 63–72.

Michael, A. 2000. Anwendung des physikalisch begründeten Erosionsprognosemodells EROSION 2D/3D—Empirische Ansätze zur Ableitung der Modellparameter. Ph.D thesis TU Freiberg.

Morgan, R.P.C., J.N. Quinton and R.J. Rickson. 1992. EUROSEM: Documentation manual (Silsoe College). Silsoe, UK.

Schindewolf, M. and W. Schmidt. 2010. Flächendeckende Abbildung der Bodenerosion durch Wasser für Sachsen unter Anwendung des Modells EROSION 3D. Schriftenreihe des Landesamtes für Umwelt, Landwirtschaft und Geologie, Heft 9/2010, Dresden. ISSN: 1867-2868.

Schindewolf, M., C. Bornkampf, M.V. Werner and J. Schmidt. 2015. Simulation of reservoir siltation with a process-based soil loss and deposition model. pp. 41–57. DOI: 10.5772/61576.

Schmidt, J. 1988. Wasserhaushalt und Feststofftransport an geneigten, landwirtschaftlich bearbeiteten Nutzflächen. Ph.D. Thesis, Free University, Dresden.

Schmidt, J. 1991. A mathematical model to simulate rainfall erosion. *In*: Bork, H.R., D.E. Ploey, A.P. Schick (eds.). Erosion, Transport and Deposition Processes—Theory and Models. Catena (Suppl.) 19: 101–109.

Schmidt, J. 1992. Modeling long-term soil loss and landform change. *In*: Abrahams, A.J. and A.D. Parsons (eds.). Overland Flow—Hydraulics and Erosion Mechanics. University College London Press, London.

Schmidt, J. 1996. Entwicklung und Anwendung eines physikalisch begründeten Simulationsmodells für die Erosion geneigter landwirtschaftlicher Nutzflächen. Berliner Geographische Abhandlungen, H. 61.

Schmidt, J., M.V. Werner and A. Routschek. 1999. Application of the EROSION-3D model to the CATSOP watershed, the Netherlands. Catena 37: 449–56.

Schob, A., J. Schmidt and R. Tenholtern. 2006. Derivation of site-related measures to minimize soil erosion on the watershed scale in the Saxonian loess belt using the model EROSION-3D. Catena 68: 153–60.

Schramm, M. 1992. Bestimmung des Bodenabtrags und des Stoffaustrags im Vorfluter eines kleinen ländlichen Einzugsgebietes. *In*: PLATE, E.J. Prognosemodell für die Gewässerbelastung durch Stofftransport aus einem kleinen ländlichen Einzugsgebiet. Schlußbericht BMFT-Verbundprojekt, S. 229–256.

Schramm, M. 1994. Ein Erosionsmodell mit räumlich und zeitlich veränderlicher Rillenmorphologie. Dissertation TH Karlsruhe.

Schwertmann, U., W. Vogland M. Kainz. 1990. Bodenerosion durch Wasser. Vorhersage des Abtrags und Bewertung von Gegenmaßnahmen, 2. Aufl., Stuttgart.

Seidel, N. and J. Schmidt. 2009. Effects of land use on surface runoff – simulations with the EROSION 3D Computer Model. Geo-öko 29 (3–4), Göttingen.

Smith, R.E. 1988. OPUS-An advanced simulation model for non-point source pollution transport at the field scale. Draft. USDA-ARS, Ft. Collins, CO.

Starkloff, T. 2012. Modelling of soil erosion with the EROSION-3D model for the Skuterud catchment in Norway. Diploma Dissertation. Soil and Water Conservation Unit, Technical University of Freiberg, Freiberg, Germany.

Van Genuchten, M. 1980. A closed-form equation for predicting the hydraulic conductivity of unsaturated soils. Soil Sci. Soc. Am. J. 44: 892–898.

Von Werner, M. 1991. Anwendung des Bodenerosionsmodells "EROSION-2D" in der agraren Nutzungsplanung. Diplomarbeit FU Berlin.

Von Werner, M. 1995. GIS-orientierte Methoden der digitalen Reliefanalyse zur Modellierung von Bodenerosion in kleinen Einzugsgebieten. Ph.D. Thesis Freie Universität, Berlin.

Weigert, A. and J. Schmidt. 2005. Water transport under winter conditions. Catena 64: 193–208.

Wischmeier, W.H. and D.D. Smith. 1965. Predicting rainfall-erosion losses from cropland east of the Rocky Mountains. Agr. Handbook 282 (USDA), Washington D.C.

Woolhiser, D.A., R.E. Smith and D.C. Goodrich. 1990. KINEROS, a kinematic runoff and erosion model: documentation and user manual. USDA–ARS–77.

Yalin, M.S. 1963. An expression for bed-load transportation. J. Hydraul. Division. Proc. Amer. Soc. of Civil Eng. 89: 221–250.

Yoon, Y.N. and H.G. Wenzel. 1971. Mechanics of sheet flow under simulated rainfall. J. Hydraulics Division Proc. Amer. Soc. Civil. Eng., 97, HY9, S. 1367–1386.

Chapter 8

SWAT Model and its Application

VK Bhatt* and AK Tiwari

Introduction

There is need to understand the physical process of erosion in relation of topography, land use and management in order to derive with best management practices. Planned land use and conservation measures to optimize the use of land and water resources help in increasing sustainable agricultural production. However, to achieve this, quantification of runoff and soil loss from the watersheds is must. Since it is very often impractical or impossible to directly measure soil loss on every piece of land, and the reliable estimates of the various hydrological parameters including runoff and soil loss for remote and inaccessible areas are tedious and time consuming by conventional methods. Therefore, it is desirable that some suitable methods and techniques are evolved for quantifying the hydrological parameters from all parts of the watersheds. Use of mathematical hydrological models to quantify runoff and soil loss for designing and evaluating alternate land use and best management practices in a watershed is one of the most viable options.

Indian Institute of Soil and Water Conservation, Research Centre, Chandigarh, 160019, India.
  Email: gmaruntiwari@gmail.com
* Corresponding author: v_k_bhatt2001@yahoo.co.in

Erosion Models for Estimating Soil Erosion

Erosion models are used to predict soil erosion. Soil erosion modeling is able to consider many of the complex interactions that influence rates of erosion by simulating erosion processes in the watershed. Various parametric models such as empirical (statistical/metric), conceptual (semi-empirical) and physical process based (deterministic) models are available to compute soil loss. In general, these models are categorized depending on the physical processes simulated by the model, the model algorithms describing these processes and the data dependence of the model. Empirical models are generally the simplest of all three model types. They are statistical in nature and based primarily on the analysis of observations and seek to characterize response from these data. The data requirements for such models are usually less as compared to conceptual and physical based models. Conceptual models play an intermediary role between empirical and physics-based models. Physical process-based models take into account the combination of the individual components that affect erosion, including the complex interactions between various factors and their spatial and temporal variabilities. These models are comparatively over-parameterized.

There have been several hydrological models developed to estimate runoff and soil loss from a watershed. USLE, RUSLE, EPIC, ANSWERS, DREAMS, CORINE, ICONA, MIKE SHE, Erosion-3D, AGNPS, CREAMS, SWAT and WEPP are few among the models. One of the major problems in testing these models is the generation of input data, that too spatially. The conventional methods proved to be too costly and time consuming for generating this input data. With the advent of remote sensing technology, deriving the spatial information on input parameters has become more handy and cost-effective. Besides with the powerful spatial processing capabilities of the Geographic Information System (GIS) and its compatibility with remote sensing data, the soil erosion modeling approaches have become more comprehensive and robust. Satellite data can be used for studying erosional features, such as gullies, rainfall interception by vegetation and vegetation cover factor. DEM (Digital Elevation Model), one of the vital inputs required for soil erosion modeling can be created by analysis of stereoscopic optical and microwave (SAR) remote sensing data. The integrated use of remote sensing and GIS could help to assess quantitative soil loss at various scales and also to identify areas that are at potential risk of soil erosion. This chapter presents the application of different hydrological models, remote sensing and GIS in estimating runoff and soil loss from the *Shivalik* foothills.

SWAT Model

Soil and Water Assessment Tool (SWAT) is a river basin model being used since 1993 mainly by hydrologists for watershed hydrology related issues (Santhi et al. 2001, Cao et al. 2006, Schuol and Ambaspour 2007, Keshta et al. 2009, Akiner and

Akkoyunlu 2012, Kushwaha and Jain 2013, Bhatt et al. 2016, Hallouz et al. 2018). It is currently one of the world's leading spatially distributed hydrological models. SWAT as a distributed parameter continuous time model was developed originally by the USDA Agricultural Research Service (ARS) and Texas A&M University (Arnold et al. 1998). It divides a watershed into smaller discrete calculation units for which the spatial variation of the major physical properties is limited, and hydrological processes can be treated as being homogeneous. The total watershed behaviour is a net result of several small sub-basins. The soil map and land user map within sub-basin boundaries are used to generate a homogeneous physical property, i.e., Hydrological Response Unit (HRU). The water balance for HRUs is computed on a daily time step. Hence, SWAT subdivides the river basin into units that have similar characteristics in soil and land cover and that are located in the same sub-basin. The SWAT model has been tested for predicting runoff and soil loss throughout the world under different conditions (Abbaspour et al. 2015, Gamvroudis et al. 2015, Gyamfi et al. 2016, Anaba et al. 2017). It has emerged as one of the most widely used water quality watershed and river basin-scale models worldwide, applied extensively for a broad range of hydrologic and/ or environmental problems. The international use of SWAT can be attributed to its flexibility in addressing water resource problems, extensive networking via dozens of training workshops and the several international conferences that have been held during the past decade, comprehensive online documentation and supporting software, and an open source code that can be adapted by model users for specific application needs (Gassman et al. 2014).

However, application of SWAT for prediction of runoff from micro-watershed is limited. Data of several years is required for development of a long term plan for homogeneous watersheds. The hydrologic component of SWAT is based on the following water balance equation:

$$SW_t = SW + \sum_{i=1}^{t} (R_i - Q_i - ET_i - P_i - QR_i) \qquad (1)$$

where: SWt is the final soil water content (mm), SW is the water content available for plant uptake, defined as the initial soil water content minus the permanent wilting point water content (mm), t is time in days, R is rainfall (mm), Qi is surface runoff (mm), ETi is evapotranspiration (mm), Pi is percolation (mm) and QRi is return flow. SWAT incorporates some of the most common hydrological equations for the simulation of flow. For the accurate implementation of these equations, detailed input data are needed. The digital elevation model (DEM) of the watershed, the soil and land use data and the climatic data of the area are required input to the model. The importance of land uses in the operation of the model lies mainly in the computation of surface runoff with the help of the SCS curve. The model includes in its database 102 different land use types, with each one assigned to a CN-II value (Curve Number for hydrological condition II). The user is required to link each of the land uses that appear in the watershed, to the ones that the model can identify. The success of the simulation depends highly on the accuracy of soils and land uses.

In most of the developing countries of the world, the majority of the basins are either sparsely gauged or not gauged at all. This necessitates the application of a robust model for estimation of runoff and sediment. In the present study the Arc View-SWAT interface (AVSWAT-X version 2005) was used to delineate the watershed boundary and the burning option to derive drainage network of choe gauging watershed. The main objective was to evaluate the applicability and performance of the model in predicting yearly water yield. In order to achieve the objectives sensitivity analysis, calibration and validation of the model were essential steps for model testing as well as extending the application area.

Evaluation of SWAT Model in Lower Himalayas

Study area

The Soil and Water Assessment Tool (SWAT) was applied to the forest micro-watershed of the lower Himalayan region. This watershed is named as Choe gauging watershed and has an area of 21 ha. The watershed is located at an altitude of 34.74°N and longitude of 76.86°E. It is located at an elevation of 350 m. Land use of watershed comprises deciduous forest, range forest and water. Nine years (1973–81) monthly data was taken up to model the hydrological output. The model was calibrated for the period 1973–1978 using the parameters based on sensitivity analysis and validated for the period 1979–1981. Loamy sand is the dominant soil type in the area. Meteorological station is located near the watershed.

Materials and methods

The Arc View-SWAT interface (AVSWAT-X version 2005) was used to delineate the watershed boundary and the burning option to derive drainage pattern of the watershed. The multiple Hydrological Response Unit (HRU) option available in AVSWAT-X interface was used with the objective to represent each field as a separate HRU. SWAT is a physically based hydrologic model which requires physically based data. Obtaining physical based data for hydrological modeling is often difficult, even in developed countries where data of high quality are generally collected and analyzed. In this study input data was collected from various sources. Meteorological input data include daily precipitation, maximum and minimum air temperature, wind speed and relative humidity.

Digital Elevation Model

The DEM data was derived from TINS created from contours of the spot elevations surveyed for the whole watersheds. The drainage Networks traced using the GPS was merged and burnt on the DEM data to exactly align the watershed outlets. The salient morphological features of the study watersheds are as shown in Table 1.

DEM and subwatersheds with HRUs are shown in Fig. 1. Land use and soil data were incorporated into SWAT model and used for reclassifying the land use and soil data. Various GIS data preprocessor modules which involve watershed delineation, input map characterization and processing, stream and outlet definition, the computation of the geomorphic parameters, and characterization of the land use/land cover and soil were developed in the course of modeling the catchment. The simulation option of the rainfall runoff modeling was performed on the basis of previous modeling techniques. These include using a curve number method for calculating surface runoff (USDA-SCS 1972), a first order Marcov Chain Skewed Normal to determine rainfall distribution, computing potential evaporation by using Penman Monteith method and Muskingum routing method for routing water through the channel networks.

Table 1. Morphological characteristics of Choe gauging micro-watersheds.

Watershed Parameters	Values
Area, ha	21
Length width ratio	1.33
Length of main drain, m	1080
Average slope, %	10
Relief, m	81
Time of concentration, min	11.43

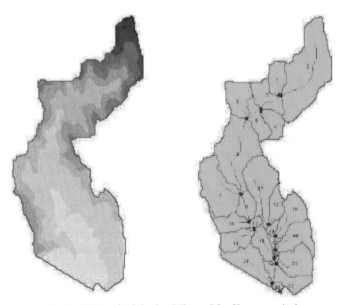

Fig. 1. DEM and sub-basins delineated for Choe watershed.

Land use data

The predominant land use and land of the micro watershed are range brush and forest respectively and the details were derived using the Google earth high resolution imageries and GPS reconnaissance in the watersheds. The classification schemes were adopted using the threshold visual color separation techniques identical to color signatures of the land use codified using the GPS data of the geo rectified Google earth images (jpeg images) captured from the GE interface. The classification was carried out using ERDAS imagine corrected to merge the unclassified cells into the nearest neighborhood cells. The classified land use map details are shown in Table 2.

Table 2. Location and land use details of micro-watershed.

Micro-watershed	Latitude-N	Longitude-E	Elevation, m	Land use	SWAT code	% area
Choe gauging, near Chandigarh	34.74	76.86	350	• Water	WATR	1.94
				• Mixed forest	FRST	27.48
				• Range brush	RNGB	70.58

After simulation process sensitivity analysis was performed involving the parameters, deep aquifer percolation (rchrg_dp), minimum depth of water in soil for base flow to occur (GWQ_MIN), initial SCS-CN for AMC-II (CN2), available water capacity of soil layer (SOL_AWC), etc., manually and by auto-calibration using SWAT CUP software version 3.1.3. Curve number and base flow parameter (Alpha_BF and Alpha_BNK) were found most sensitive parameter for the study watershed. Ranges of parameters used for modeling are shown in Table 3. Calibration and validation were performed for the period 1973–78 and 1979–81 respectively.

SWAT-CUP 2009 version was used to calibrate the model using Sequential uncertainty fitting (SUFI ver2). SUFI-2 is one of five different modules (SUFI2, ParaSol, GLUE, MCMC and PSO) that are linked with SWAT in the package called SWAT Calibration Uncertainty Programs (SWAT-CUP). Its main function is to calibrate SWAT and perform validation, sensitivity and uncertainty analysis for a watershed model created by SWAT.

Various SWAT parameters for estimation discharge were estimated using the SUFI-2 program (Abbaspour et al. 2007). Uncertainty is defined as discrepancy between observed and simulated variables in SUFI-2 where it is counted by variation between them. SUFI-2 combines calibration and uncertainty analysis to find parameter uncertainties while calculating smallest possible prediction uncertainty band. Hence, these parameters uncertainty reflect all sources of uncertainty, i.e., conceptual model, forcing inputs (e.g., temperature) and the parameters themselves. In SUFI-2, uncertainty of input parameters is depicted

Table 3. Range of parameter values used for modeling of Choe gauging watershed.

Sr. No.	Parameter code	Description	Fitted value	Min. value	Max. value	Location
1	SOL_K	Saturated Hydraulic Conductivity (mm/hr)	0.8957	0.64	0.95	*.sol
2	ALPHA_BNK	Baseflow alpha factor for bank storage	0.2707	0.17	0.30	*.rte
3	SOL_BD	Soil Bulk density	0.4187	0.34	0.43	*.sol
4	ESCO	Soil Evaporation Compesation Factor	0.0412	.03	0.06	*.bsn
5	SLSUBBSN	Average slope length of basin	0.2550	0.18	0.30	*.hru
6	CN2	SCS Runoff Curve Number	0.4025	0.37	0.47	*.mgt
7	REVAPMIN	Threshold depth of water in the shallow aquifer for"revap" to occur	1.6575	1.50	2.20	*.gw
8	RCHRG_DP	Deep Aquifer Percolation Factor	0.6087	0.54	0.65	*.gw
9	CH_K2	Effective hydraulic conductivity of main channel	193.725	150.0	203.0	*.rte
10	GW_DELAY	Ground Water Delay (days)	63.025	61.0	70.0	*.gw
11	GWQMN	Theshold depth of water in the shallow aquifer for flow to occur (mm)	2128.75	1980.0	2150.0	*.gw
12	GW_REVAP	Ground Water "Revap" Coefficient	2.545	2.50	3.10	*.gw

as a uniform distribution, while model output uncertainty is quantified at the 95% prediction of uncertainty (95PPU). The cumulative distribution of an output variable is obtained through Latin hypercube sampling.

SUFI-2 starts by assuming a large parameter uncertainty within a physically meaningful range, so that the measured data fall initially within 95PPU, then narrows this uncertainty in steps while monitoring the P_factor and R_factor. The P_factor is the percentage of data bracketed by 95% prediction uncertainty (95PPU) and R_factor is the ratio of average thickness of the 95PPU band to the standard deviation of the corresponding measured variable. A p-factor of 1 and R-factor of zero is a simulation that exactly corresponded to measured data. In each iteration, previous parameter ranges are updated by calculating the sensitivity matrix and the equivalent of a Hessian matrix, followed by the calculation matrix. Parameters are then updated in such a way that new ranges are always smaller than previous ranges and are centred on the best simulation (Abbaspor et al. 2007). These two measured factors can be used as statistical analysis instead of the usual equations such as coefficient of determination (R^2), Nash-Sutcliffe (Nash and Sutcliffe 1970) which only compares two signals. Other statistical analyses in this study are the coefficient of determination R^2 multiplied by the coefficient of the regression line and Root Mean Square Error (RMSE).

The objective function used to test the model performance were the Nash and Sutcliffe model efficiency (η) and Root Mean Square Error (RMSE). These functions are given as follows:

$$\eta = ((FIV\text{-}FRV)/FIV)*100 \qquad (2)$$

$$\text{Where, FIV} = \text{Initial variance} = \sum_{i=1}^{n}(Q_i - \bar{Q})^2 \qquad (3)$$

$$\text{FRV} = \text{Remaining variance} = \sum_{i=1}^{n}(Q_i - \hat{Q}_i)^2 \qquad (4)$$

$$\text{RMSE} = (\sum_{i=1}^{n}(Q_i - \hat{Q}_i)^2/n)^{1/2} \qquad (5)$$

Where Q_i and $\hat{Q}_i$ are i^{th} observed and computed values of the rainfall series, $\bar{Q}$ is mean of observed rainfall series and n is length of data.

Results and discussion

Month wise simulation was carried out for both watersheds for different periods. Calibration and validation of the SWAT model output with observed values are shown for the watershed in Fig. 2 and Fig. 3. For Choe gauging watershed, monthly observed data of the monsoon season from 1973 to 1977 was taken for calibration and three years data from 1978 to 1981 was taken for validation of model. Calibrated and validated SWAT model outputs were found to be reasonably

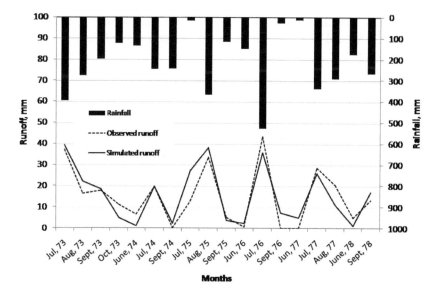

Fig. 2. Simulated and observed runoff for Calibration period.

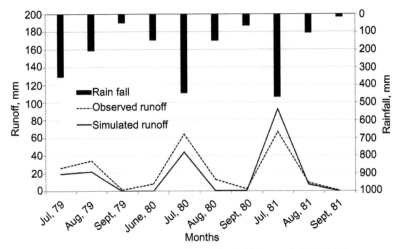

Fig. 3. Simulated and observed runoff for Validation period (1979–81).

Table 4. Performance evaluation of model for calibration and validation.

Performance measure	Calibration	Validation
Nash and Sutcliffe Efficiency, %	80.2	73.2
RMSE	5.8	12.3
R^2	0.81	0.83

simulating the observed runoff values. Nash and Sutcliffe Efficiency were found to be 80.2% for calibration and 73.3% (Table 4) for validation, as such these values can be considered reasonably well for any model. Similarly, RMSE was calculated as 5.8 and 12.3 respectively for calibration and validation.

Conclusions

Although the model generates detailed outputs at the spatial and temporal scale, in the present analyses only water yield has been considered and reported. The application of the SWAT model in generation of crucial information such as water and sediment can be used.

The main objective of the study was to evaluate the performance and applicability of the SWAT model in predicting the hydrology of the micro-watershed of the lower Himalayas. The ability of this model to predict surface runoff was evaluated through sensitivity analysis, model calibration and model validation. The sensitive parameters were used to find the most reasonable parameter values for optimum estimation of runoff. The analysis shows that base flow parameter, i.e., alfa factor (days) and SCS curve number were found as the most sensitive parameter for both the watersheds.

Model performance evaluation statistics for simulating monthly runoff for Choe gauging watershed, both for calibration and validation periods showed that simulated runoff matched well with the observed data. The study has shown that the SWAT model can produce reliable estimate of monthly runoff even for micro watersheds. Thus, SWAT is a good modeling tool for analysis of hydrological processes and water resource planning.

References

Abbaspour, K.C., J. Yang, L. Maximov, R. Siber and K. Bogner. 2007. Modeling hydrology and water quality in the pre-alpine/alpine Thur watershed using SWAT. J. Hydrol. 333: 413–430.

Abbaspour, K.C., E. Rouholahnejad, S. Vaghefi, R. Srinivasan, H. Yang and B. Klove. 2015. A continental-scale hydrology and water quality model for Europe: Calibration and uncertainty of a high-resolution large-scale SWAT Model. J. Hydrol. 524: 733–752. https://doi.org/10.1016/j.jhydrol.2015.03.027.

Anaba, L.A., N. Banadda, N. Kiggundu, J. Wanyama, B. Engel and D. Moriasi. 2017. Application of SWAT to assess the effects of land use change in the Murchison bay catchment in Uganda. Computat. WaterEnergyEnviron. Engg. 6: 24–40.

Akiner, M.E. and A. Akkoyunlu. 2012. Modeling and forecasting river flow rate from the Melen Watershed, Turkey. J. Hydrol. 456-457: 121–129.

Arnold, J.G., R. Srinivasan, R.S. Muttiah and J.R. William. 1998. Large area hydrological modeling and assessment: Part I: Model Development. J. Am. Water Resour. As. 34: 73–89.

Bhatt, V.K., A.K. Tiwari and D.R. Sena. 2016. Application of SWAT model for simulation of runoff in micro watersheds of lower Himalayan region of India. Indian J. Soil Conserv. 44: 133–140.

Cao, W., B.W. Bowden and T. Davie. 2006. Multi-variable and multisite calibration and validation of SWAT in a large mountainous watershed with high spatial variability. Hydrol. Process. 20: 1057–1073.

Gamvroudis, C., N.P. Nikolaidis, O. Tzoraki, V. Papadoulakis and N. Karalemas. 2015. Water and sediment transport modeling of a large temporary river basin in Greece. Sci. Total Environ. 508: 354–365. https://doi.org/10.1016/j.scitotenv.2014.12.005.

Gassman, P., A.M. Sadhegi and R. Srinivasan. 2014. Applications of the SWAT model-special section: overview and insights. J. Environ. Qual. 43: 1–8. DOI: 10.2134/jeq2013.11.0466.

Gyamfi, C., J.M. Ndambuki and R.W. Salim. 2016. Application of SWAT Model to the Olifants Basin: Calibration, Validation and Uncertainty Analysis. J. Water Resour. Prot. 8: 397–410. http://dx.doi.org/10.4236/jwarp.2016.83033.

Hallouz, F., M. Meddi, G. Mahe, S. Alirahmani and A. Keddar. 2018. Modeling of discharge and sediment transport through the SWAT model in the basin of Harraza (Northwest of Algeria). Water Sci. 32: 79–88.

Keshta, N., A. Elshorbagy and S. Carey 2009. A generic system dynamics model for simulating and evaluating the hydrological performance of reconstructed watersheds. Hydrol. Earth Syst. Sci. 13: 865–881.

Kushwaha, A. and M. Jain. 2013. Hydrological simulation in a forest dominated watershed in Himalayan Region using SWAT Model. Water Res. Manage. 27: 3005–3023.

Nash, J.E. and J.V. Sutcliffe. 1970. River flow forecasting through conceptual models. J. Hydrol. 10: 282–290.

Santhi, C., J.G. Arnold, J.R. Williams, W.A. Dugas and L. Hauck. 2001. Validation of the SWAT model on a large River basin with point and nonpoint sources. J. Am. Water Resour. As. 37: 1169–1188.

Schuol, J. and K.C. Abbaspour. 2007. Using monthly weather statistics to generate daily data in a SWAT model application to West Africa. Ecol. Model. 201: 301–311.

USDA-SCS. 1972. National Engineering Handbook Section 4, Hydrology. USDA-SCS, Washington, DC, USA.

Chapter 9

Watershed Management in the 21st Century

Seyed Hamidreza Sadeghi

Why Watershed Management?

Increasing land degradation in the world with degradation moving at a quicker speed in the developing countries has become a serious problem threatening soil and water resources (Bartarya 1991, Sidle 2000, Sadeghi et al. 2004). During the last few decades, natural resources worldwide have faced some serious degradation problems such as soil erosion, sedimentation, wind erosion, water scarcity and pollution, groundwater overexploitation, land use changes, overgrazing in the rangelands, soil salinity, forest fire, flooding and wetlands loss (Bartarya 1991, Sidle 2000). Finding scientifically appropriate, practically feasible, environmentally friendly, technically sound, economically efficient, developmentally sustainable and socially acceptable solutions are therefore vital for the successful and persistent management of diminishing resources.

"A watershed is a complex and dynamic bio-physical system which is identified as planning and management unit. Hence, considering all technical, socio-economical, physical, ecological and organizational dimensions is essential for proper planning and management processes. Due to complex interactions among different aspects of the watershed, application of an integrated management approach is inevitable to coordinate study aspects" (California Department

Department of Watershed Management Engineering, Faculty of Natural Resources, Member of Agrohydrology Group, Tarbiat Modares University, Noor 4641776489, Mazandaran Province, and President of Watershed Management Society of Iran, Iran.
Email: sadeghi@modares.ac.ir

of Conservation 2015). A watershed is also a hydrological and biophysical response unit, and a holistic ecosystem in terms of the materials, energy, and information present. The watershed not only is a useful unit for physical analyses, it can also be a suitable socioeconomic-political unit for management planning and implementation. In essence, a watershed is a basic organizing unit to manage resources. Watershed management is faced with complex problems that are characterized by uncertainty and change. Watershed management is an ever-evolving practice involving the management of land, water, biota, and other resources in a defined area for ecological, social, and economic purposes (Wang et al. 2016). It studies the relevant characteristics of a watershed aimed at the sustainable distribution of its resources and the process of creating and implementing plans, programs, and projects to sustain and enhance watershed functions affecting the plant, animal, and human communities within the watershed boundary (California Department of Conservation 2015).

In 2015, some 17 Sustainable Development Goals (www.un.org) were designated to be achieved as the 2030 Agenda for Sustainable Development none of which can be adopted without an integrated and adaptive watershed management. This can be planned in different zones of the watersheds namely headwaters (upstream), transfer zone (midstream) and depositional zone (downstream) as an occurrence that changes the pattern of all that follows, moving the flow of events toward a different outcome and simply called the "watershed event" (California Department of Conservation 2015). This is in the same vein as the United Nations; which has committed to focus on water for a decade (2018–2028) to advance sustainable development of water (www.un.org). Integrated and systematic management of the watershed is one of the vital approaches to develop sustainably (Sadoddin et al. 2016, Raum 2018). It leads to the effective utilization of natural resources, alleviates poverty, improves sustainable livelihoods, and increases collaboration among the various stakeholders particularly in undeveloped countries.

To sustainably utilize available resources and meet the fulfillment of human needs as well as restore the ecosystem balance to mitigate poverty, conserve the Earth and ascertain prosperity for all as part of a new sustainable development agenda in 21st century, new breath of air has to be blown into existing programs and projects at the watershed scale. Otherwise, all accessible and even available resources will have thoroughly perished resulting in an irreversible situation, including tragic famine and cannibalism at the end. This problem is far more serious in developing countries where higher demands exist on one side and low technology and technical knowledge are available (World Bank 2008) on the other side. To address these issues, some important and developing strategies have been discussed for the integrated management of the watershed through incorporating various techniques and using indigenous knowledge.

In summary, integrated watershed management is the process of creating and implementing plans, programs, and projects to sustain and enhance watershed functions that provide the goods, services, and values desired by the community affected by the conditions within a watershed boundary. The management is integrated and complex, including components inside (e.g., upstream, midstream, downstream) and outside the watershed, affecting both man-made and natural factors. There is a general belief that the future watershed management will need to account for the management of limited resources for ever-increasing demand and greed of the humans by employing technological advancements and holistic, cross-disciplinary approaches (Wang et al. 2016). It certainly ensures for watersheds continue to efficiently and successfully perform essential ecosystem services and serve their ecological and socio-economic services.

Optimizing Land Use Utilization

Considering conflicts upon very limited resources among different sectors with an ever-increasing demand at the watershed scale is necessary in the 21st century for satisfying the inhabitant's demand (Pastori et al. 2017, Tajbakhsh et al. 2018, Wu 2018). Otherwise, more transient benefits reach a particular sector resulting in the diminishing of other particles of the ecosystem and ultimately making it vulnerable to introducing driving forces. It virtually leads to degradation of the watershed and the consequent lower productivity, and further subsequent overexploitation of the resources to compensate insufficient production. Optimization as an act, process, or methodology of the decision making process as fully, perfect, functional and efficient as possible is therefore needed for effectively managing the valuable watershed assets and improving the ecosystem services (Guo et al. 2016, Li and Ma 2017).

Despite the plethora of literature existing on the optimization field of different resources management (Sadeghi et al. 2009, Liu et al. 2015, Ma and Zhao 2015, Pastori et al. 2015, Tajbakhsh et al. 2018, Wu 2018), frequent application of this applied mathematical approach in the management of the watershed as a necessary approach for resources management in the developing world is still lacking. However, the application of linear programming as a basic method for many other optimization programs has been reported for the optimization of watershed management using associated software. A specific study on the optimization of land use allocation to orchard, range, irrigated and dry farming land uses has been reported by Sadeghi et al. (2009) using linear programming to minimize soil erosion and to maximize the economic return within the Birmvand Watershed in Kermanshah Province, Iran. Solving the multi-objectives linear optimization problem developed for the study watershed; revealed that the amount of soil erosion and benefit could respectively reduce and increase by 7.9% and 18.6%, in the case of implementing optimal allocation of the land use in the study. It is therefore an approach that can be applicable for the resources management of the watershed (Kaim et al. 2018).

Monitoring based Watershed Management

Understanding hydrologic behaviors and assessing the influence of land use and land cover change in the hydrologic response of different watersheds is important for watershed researchers and management in the 21st century (Paule-Mercado et al. 2018). Assessment of watershed behavior is intended to provide a better understanding and awareness of hydrologic behaviors of the watershed system. So that monitoring the health of watersheds is a critical precursor to the adaptive resources management on a watershed basis (Hazbavi et al. 2018a and b). Hence, continuous monitoring of the watersheds behavior in response to various anthropogenic and natural driving triggers helps adopt an integrated and adaptive management strategy in different scales (Hazbavi and Sadeghi 2017, Hazbavi et al. 2018a and b). Monitoring of the watershed health is hence considered to be one of the main stages of adaptive management and the main components of the watershed management plan as shown in Fig. 1.

Developing better methods for analyzing and assessing cumulative watershed effects is considered to be the other challenges in the 21st century (Sidle 2000). That is why; ascription of anthropogenic and natural effects on watershed degradation is also needed.

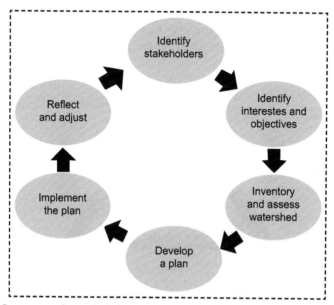

Fig. 1. Components of watershed adaptive management plan (After www.conservation.ca.gov).

Adaptive Watershed Management

Concepts and definition

Despite existence of the concept of watershed management for millennia (Wang et al. 2016), the management of watershed resources has not been successfully achieved. Such that almost half of the countries in the world have low to very low access to fresh water; owing in part to population growth resulting in increasing constraints on the land, water, and other natural resources availability. Therefore, the scarcity of fresh water supply, contamination of agricultural land, and polluted streams are affecting millions of lives (Wang et al. 2016). It clearly verifies the necessity of a new definition for watershed management to comprehensively and suitably satisfy man's ever-increasing demands, and appropriately and sustainably conserve the environment. Such a definition will certainly be fulfilled using a system's approach to sustainable development as proposed by Flint in 2015 (www.eeeee.net/watershed.htm). Therefore there should be a heartfelt belief that the current policies affecting civilization today and the Earth for generations yet to come. There should be another belief that we cannot independently alter or modify one element of a natural system without expecting changes elsewhere. We therefore need to think like a watershed because acting sustainably requires concurrent multi-dimensional thinking in such way to cover both the temporal and spatial for various sectors concerning the economy, society and environment. Accordingly an adaptive watershed management is a decision-making process which effectively integrates both short and long-term economic, environmental and social concerns (Flint 2006). Adaptive watershed management is cored on *Five E's* including Ecology, Economy, Equity, Education and Evaluation as shown in Fig. 2; leading and encouraging the development of interdisciplinary

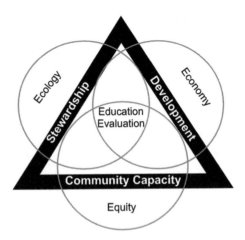

Fig. 2. *Five E' Unlimited* in watershed management (After www.eeeee.net/watershed.htm).

adaptive, and integrally-informed understanding of social-ecological systems at the watershed level. The knowledge is then transferred to various stakeholders to facilitate collaboration among the different stakeholders and stewards.

Adaptive management can be amalgamated with integrated and comprehensive management helping to cope with the uncertainties dealing with the management of complicated and dynamic systems of the watershed. Basically, adaptive management gives several suggestions for handling of the governing complexity on the watershed system. It is achieved by learning from the watershed outcomes while doing and working with the system. It ultimately leads to a proper adoption of managerial approaches and adjustment of future strategies accordingly. In order to get access to a successful adaptive management, an insight goal based monitoring and consequent evaluations are needed. Forthcoming strategies for the adaptive management of the watershed need to be persistently improved by learning from the pre-implemented policies. Self-evaluation to identify mistakes, flexibility in decisions, appropriate regulations to rectify mistakes, time dedication and financial investments in reducing the biases from the main goals are therefore a necessary need for the adaptive watershed management (Allan et al. 2008, Raadgever et al. 2008, Porzecanski et al. 2012).

Low impact development

Low impact development practices as a cost effective practice of land development approach strives to mimic the pre-development conditions of a watershed (Ahiablame and Shakya 2016, Xu et al. 2018) and can be used to mitigate risks in watersheds, adaptively. The low impact development strategy has attracted growing attention as an important, efficient and more reliable method for the management of urban watersheds with a focus on flood mitigation (Ahiablame and Shakya 2016, Hu et al. 2017, Xu et al. 2018). The findings verified effective roles of low impact development practices for flood inundation mitigation at the watershed scale. Basically, the low impact development approach for storm water control is shaped based on the eight main elements represented in Fig. 3.

Water-Energy-Food nexus

To adaptively manage the watersheds, other new approaches such as the Water-Energy-Food (WEF) nexus can also be adopted, since the objective of watershed management can be achieved by the application of interdisciplinary and professional approaches through establishing a dynamic and optimal balance in supply and demand resources and consumption. The WEF nexus has been initially introduced in the world as an adaptive management approach to reduce the vulnerability to climate change and human impact in terms of the security

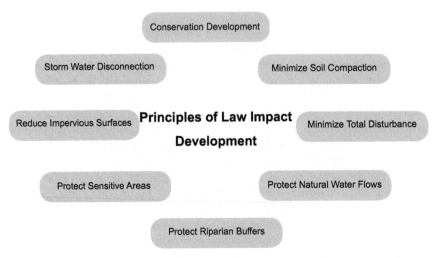

Fig. 3. Main principles of low impact development measures for storm water management (After Vermont Green Infrastructure Initiative 2018).

challenges of water, energy, and food (Endo et al. 2015, Rasul and Sharma 2016). The WEF nexus focuses on the interdependency of water, energy, and food security to be explicitly identified in the decision making process (Mohtar et al. 2015). By definition, the nexus consists of basic concepts for the dynamics of the water, energy and food inter-relationship (Smajgl et al. 2016) water or food security. Current frameworks are partial as they largely represent a water-centric perspective. Our hypothesis is that a dynamic nexus framework that attempts to equally weight sectoral objectives provides a new paradigm for diagnosis and investigation. Dynamic refers here to explicitly understanding or a diagnosis of an important discussion in agricultural land use in watersheds, presents several challenges within the WEF nexus at the local and global scales (Gulati and Pahuja 2015). While considering other important chapters of the watershed system such as soil there is a need to allow the comprehensive management of the watershed in an adaptive manner (Lal et al. 2017).

Actually, to consider the nexus approach with a sufficient concentration on the soil is essential as the foundation of the future of mankind. It is therefore a long way ahead in the future to provide meaningful concepts of the WEF nexus at the watershed scale. It is due to the complexities in the dynamic components of the watershed system. In light of the evidence, most of the literature on WEF and its various editions exist in Central, South, Southeast and East at 38%; and North America (USA, Mexico and Canada) with another 31%. It clearly verifies that more necessary efforts need to be made in other regions with a further focus in the developing countries where such approaches are needed to harmonize the inter-relationship amongst the important chapters of soil, water, energy and food.

Best co-management of the watershed

Conservation of available and accessible resources and making a balance amongst the ecosystem components is essential to target the goals of the developmental plan. Hence, it is designed in order to recognize that the natural resources are the mutual assets of the society. If such an approach is governed, all different categories of people can receive mutual benefits and have co-responsibility for its use and management. Towards this, empowerment and deliberated participation, capability strengthening as well as awareness and encouragement are important for decision making on watershed management. It eventually leads to a balanced conservation and use, short term and long term benefits for the stakeholders. Establishing a government sector, non-government organization, local community and academia network is also supposed to be an actual social driving force (Chanya et al. 2014).

Best management practices (BMPs) are famous and effective approaches applied to ameliorate hydrological and fluvial behaviors of the watersheds (Strauch et al. 2013, Loperfido et al. 2014). Despite the long-term performance of the BMPs, they can help improve the decision support systems for creating competent strategies for watershed management projects (Liu et al. 2018) in sustainable and adaptive manner. BMPs widely remain an important solution at the local and regional-scale to mitigate water quantity and quality issues. Such that, the BMPs tries to handle issues in a very soft manner with the least intervention in the natural system. In particular, the watershed-wide application of the distributed BMPs improved hydrological behavior of the system. Integrated planning of storm water management system, protected riparian buffers and forest land cover with suburban development in the distributed-BMP watershed which enabled multi-purpose use of land, that provided an aesthetic value and green-space, community gathering points, and wildlife habitat in addition to an individual hydrologic storm water treatment (Loperfido et al. 2014).

Different features of a watershed include soil and water utilization and conservation, water rights, and the overall planning and utilization of watersheds. Many stakeholders viz. landowners, land use organizations, foresters, ranchers, environmentalists, governmental agencies, and communities all play an integral part in watershed management. Once all stockholders and individuals think like a watershed and become aware of the benefits of proper thinking, they are often hands-on in the different stages of watershed management programs. They then favorably contribute in the planning, decision-making, implementation, restoration, maintenance, monitoring and even evaluation processes. It will eventually help reduce conflicts, increase commitment to the essential actions to fulfill environmental aims, improve economic and social situations of the stockholders and ultimately lead to sustainable life and development (Flint 2006).

A new cooperative watershed management methodology is designed for developing equitable and efficient BMPs with the participation of all main stakeholders. The approach intended to control watershed outputs and to improve the socio-economic status of the watershed, considering villagers, legislation and executive stakeholders with conflicting interests (Adhami et al. 2018). Toward this goal, a powerful compromising like game theory can be used for analyzing strategies amongst the various demands in order to achieve cooperative decision-making in the sub-watershed and Best Co-Management Practices BCMPs prioritization. The collaborative management is a vast ranged-effort to participate in information collecting, decision-making and an accomplishment of projects (Bryson et al. 2013) and helps decision-makers resolve complex society-environment dilemmas (Leys and Vanclay 2011) at the watershed scale. A collaborative watershed management is a process, which includes relevant stakeholders to watershed resources in decision-making to achieve ecosystem-oriented goals, such as water quality improvement, soil conservation and pollution control (Ucler et al. 2015, Thomas 2017).

Conclusion

There are many conflicting issues dealing with watershed ecosystems in the 21st century none of which can be simply ignored. The necessity of considering the ecosystem balance on one side and the growing demands of communities as the main driving forces on the system on another side make it somehow difficult to achieve sustainable development. Adopting appropriate approaches such as adaptive management, co-best management practices and optimal scenarios may therefore be as practical approaches in the current century. These approaches hopefully guarantee the cautious utilization of the available resources in a way to restore natural potentials and conserve them for future generations. However, continuous monitoring of the watershed systems to evaluate the outcome behaviors and accordingly adapt our attitudes in the proper direction is essentially needed. Obviously, more attention and considerations along with insight investigations are required in developing countries where the degradation of various resources is drastically accelerated.

Acknowledgment

The valuable efforts of Miss. A. Katebi Kord and Mr. E. Sharifi Moghaddam for reproducing some figures as well as final reading of Dr. Z. Hazbavi are greatly appreciated.

References

Adhami, M., S.H. Sadeghi and M. Sheikhmohammady. 2018. Making competent land use policy using a co-management framework. Land Use Policy 72: 171–180.
Ahiablame, L.R. and R. Shakya. 2016. Modeling flood reduction effects of low impact development at a watershed scale. J. Environ. Manag. 171: 81–91.

Allan, C., A. Curtis, G. Stankeyand B. Shindler. 2008. Adaptive management and watersheds: a social science perspective. J. Am. Water Resour. Assoc. 44: 166–174.

Bartarya, S.K. 1991. Watershed Management Strategies in Central Himalaya: The Gaula River Basin, Kumaun, India, Land Use Policy 8: 177–184.

Bryson, J.M., K.S. Quick, C.S. Slotterback and B.C. Crosby. 2013. Designing public participation processes. Public Adm. Rev. 73: 23–34.

California Department of Conservation. 2015. Watershed Program. http://www.conservation.ca.gov/dlrp/wp/Pages/Index.aspx. Accessed June 9, 2018.

Chanya, A., B. Prachaak and T.K. Ngang. 2014. Conflict management on use of watershed resources. ProcediaSoc. Behav. Sci. 136: 481–485.

Endo, A., K. Burnett, P.M. Orencio, T. Kumazawa, C.A. Wada, A. Ishii and M. Taniguchi. 2015. Methods of the water-energy-food nexus. Water. 7: 5806–5830.

Flint, R.W. 2006. Water resources sustainable management: Thinking like a watershed. Ann. Arid Zone 45: 399–423.

Gulati, M. and S. Pahuja. 2015. Direct delivery of power subsidy to manage energy–ground water–agriculture nexus. Aquat. Procedia 5: 22–30.

Guo, X.Y., X.L. Liu and L.G. Wang. 2016. Land use optimization in order to improve ecosystem service: A case of Lanzhou City. Acta Ecol. 36: 7992–8001.

Hazbavi, Z. and S.H.R. Sadeghi. 2017. Watershed health characterization using reliability-resilience-vulnerability conceptual framework based on hydrological responses. Land Degrad. Dev. 28: 1528–1537.

Hazbavi, Z., J.E.M. Baartman, J.P. Nunes, S.D. Keesstra and S.H.R. Sadeghi. 2018a. Changeability of reliability, resilience and vulnerability indicators with respect to drought patterns. Ecol. Indic. 87: 196–208.

Hazbavi, Z., S.D. Keesstra, J.P. Nunes, J.E.M. Baartman, M. Gholamalifard and S.H.R. Sadeghi. 2018b. Health comparative comprehensive assessment of watersheds with different climates. Ecol. Indic. 93: 781–790.

https://www.un.org/sustainabledevelopment, Last access in May 19, 2018.

https://www.eeeee.net/watershed.htm, Last access in June 2, 2018.

Hu, H., T. Sayama, X. Zhang, K. Tanaka, K. Takara and H. Yang. 2017. Evaluation of low impact development approach for mitigating flood inundation at a watershed scale in China. J. Environ. Manage. 193: 430–438

Kaim, A., A.F. Cord and M. Volk. 2018. A review of multi-criteria optimization techniques for agricultural land use allocation, Environ. Model. Softw. 105: 79–93.

Lal, R., R.H. Mohtar, A.T. Assi, R. Ray, H. Baybil and M. Jahn. 2017. Soil as a basic nexus Tool: soils at the center of the food–energy–water nexus. Renew. Sustainable Energy Rev. 4: 117–129.

Leys, A.J. and J.K. Vanclay. 2011. Social learning: A knowledge and capacity building approach for adaptive co-management of contested landscapes. Land Use Policy 28: 574–584.

Li, X. and X. Ma. 2017. An uncertain programming model for land use structure optimization to promote effectiveness of land use planning. Chin. Geogr. Sci. 27: 974–988.

Liu, Y., W. Tang, J. He, Y. Liu, T. Ai and D. Liu. 2015. A land-use spatial optimization model based on genetic optimization and game theory. Comput. Environ. Urban Syst. 49: 1–14.

Liu, Y., B.A. Engel, D.C. Flanagan, M.W. Gitau, S.K. McMillan, I. Chaubey and S. Singh. 2018. Modeling framework for representing long-term effectiveness of best management practices in addressing hydrology and water quality problems: Framework development and demonstration using a Bayesian method. J. Hydrol. 560: 530–545.

Loperfido, J.V., G.B. Noe, S.T. Jarnagin and D.M. Hogan. 2014. Effects of distributed and centralized stormwater best management practices and land cover on urban stream hydrology at the catchment scale. J. Hydrol. 519: 2584–2595.

Ma, X. and X. Zhao. 2015. Land use allocation based on a multi-objective artificial immune optimization model: An application in Anlu county, China. Sustainability 7: 15632–15651.

Mohtar, R.H., A.T. Assi and B.T. Daher. 2015. Bridging the water and food gap: The role of the water-energy-food nexus. Unu-Flores 5: 1–31.

Pastori, M., A. Udías, F. Bouraoui, A. Aloe and G. Bidoglio. 2015. Multi-objective optimization for improved agricultural water and nitrogen management in selected regions of Africa. Int. Series Operat. Res. Manage. Sci. 224: 241–258.

Pastori, M., A. Udías, F. Bouraoui and G. Bidoglio. 2017. Multi-objective approach to evaluate the economic and environmental impacts of alternative water and nutrient management strategies in Africa. J. Environ. Inform. 29: 16–28.

Paule-Mercado, M.C.A., I. Salim, B.Y. Lee, S. Memon, R.U. Sajjad, C. Sukhbaatar and C.H. Lee. 2018. Monitoring and quantification of stormwater runoff from mixed land use and land cover catchment in response to land development. Ecol. Indic. 93: 1112–1125.

Porzecanski, I., L.V. Saunders and M.T. Brown. 2012. Adaptive management fitness of watersheds. Ecol. Soc. 17: 29–43.

Raadgever, G.T., E. Mostert, N. Kranz, E. Interwies and J.G. Timmerman. 2008. Assessing management regimes in transboundary river basins: do they support adaptive management? Ecol. Soc 13: Article 14.

Rasul, G. and B. Sharma. 2016. The nexus approach to water–energy–food security: an option for adaptation to climate change. Clim. Policy. 16: 682–702.

Raum, S. 2018. Reasons for Adoption and Advocacy of the Ecosystem Services Concept in UK Forestry. Ecol. Econ. 143: 47–54.

Sadeghi, S.H.R., D.A. Najafi and M. Vafakhah. 2004. Study on land use variation on soil erosion (Case study: Lenjan-e-Olya in Isfahan Province). Proceedings National Conference on Watershed Management and Water and Soil Resources, Kerman, Iran, 115–123.

Sadeghi, S.H.R., K. Jalili and D. Nikkami. 2009. Land use optimization in watershed scale. Land Use Policy. 26: 186–193.

Sadoddin, A., M. Ownegh, A. N. Nejad and S.H.R. Sadeghi. 2016. Development of a National Mega Research Project on the integrated watershed management for Iran. Environ. Resour. Res. 4: 231–238.

Sidle, R.C. 2000. Watershed Challenges for the 21st Century: A Global Perspective for Mountainous Terrain USDA Forest Service Proceedings RMRS–P–13.

Smajgl, A., J. Ward and L. Pluschke. 2016. The water-food-energy nexus-realizing a new paradigm. J. Hydrol. 533: 533–540.

Strauch, M., J.E.F.W. Lima, M. Volk, C. Lorz and F. Makeschin. 2013. The impact of Best Management Practices on simulated streamflow and sediment load in a Central Brazilian catchment. J. Environ. Manage. 127 (Suppl.): S24–S36.

Tajbakhsh, S.M., H. Memarian and A. Kheyrkhah. 2018. A GIS-based integrative approach for land use optimization in a semi-arid watershed. Glob. J. of Environ. Sci. Manage. 4: 31–46.

Thomas, A. 2017. A context-based procedure for assessing participatory schemes in environmental planning. Ecol. Econ. 132: 113–123.

Ucler, N., G.O. Engin, H.G. Köçken and M.S. Öncel. 2015. Game theory and fuzzy programming approaches for bi-objective optimization of reservoir watershed management: a case study in Namazgah reservoir. Environ. Sci. Pollut. Res. 22: 6546–6558.

Vermont Green Infrastructure Initiative. 2018. Low impact development (LID) fact sheet, LID overview.http://dec.vermont.gov/sites/dec/files/wsm/erp/docs/sw_gi_1.0_LID_series.pdf,Last access in July 7, 2018.

Wang, G., S. Mang, H. Cai, S. Liu, Z. Zhang, L. Wang and J.L. Innes. 2016. Integrated watershed management: evolution, development and emerging trends. J. For. Res. 27: 967–994.

World Bank. 2008. Global Economic Prospects 2008: Technology Diffusion in the Developing World. Washington, DC: World Bank.

Wu, H. 2018. Watershed prioritization in the upper Han River basin for soil and water conservation in the South-to-North Water Transfer Project (middle route) of China. Environ. Sci. Pollut. Res. 25: 2231–2238.

Xu, T., B.A. Engel, X. Shi, L. Leng, H. Jia, S.L. Yu and Y. Liu. 2018. Marginal-cost-based greedy strategy (MCGS): Fast and reliable optimization of low impact development (LID) layout. Sci. Total Environ. 640-641: 570–580.

Chapter 10

Bio-industrial Watershed Management

Sanjay Arora

Introduction

Soil, aqua and flora are the most fundamental natural assets for sustainable development and management (Aher et al. 2014), and hence should be handled and managed efficiently, collectively and simultaneously (Aher et al. 2014). Managing the natural resources through a sustainable approach is a coherent phenomenon in its natural region (Aher et al. 2012). In this context, the natural regions are invented to be in terms of the flow of water, which influences almost all fields of the environment, where the regions are diversified as basin, catchment, sub-catchment, macro-watershed (50,000 ha), sub-watershed (10,000–50,000 ha), milli-watershed (1,000–10,000 ha), micro watershed (100–1,000 ha), and mini watershed (1–100 ha) (Nair 2009, Aher et al. 2014). Planning and management of natural resources at the micro-level of the watershed where there is a high spatio-temporal variability in the geo-physical and socio-economic variables; particularly in the fragile arid and semi-arid tropics, is the crucial need of the hour (Aher et al. 2012). The real challenge on water resources planning at a micro-level is to assess the quantum of water demand and its availability.

Poverty mostly resides in the rain fed agriculture areas. Rain fed agriculture is mostly practiced in arid, semi-arid and sub-humid regions where rainfall is

ICAR–Central Soil Salinity Research Institute, Regional Research Station, Lucknow (U.P.), 226002, India.
Email: aroraicar@gmail.com

meagre in quantity, unpredictable in distribution and is also characterized by high inter-year variability (Venkateswarlu and RamaRao 2010). Farming systems depend upon pulses, coarse cereals, oilseeds, cotton, pearl millet, sorghum and other low moisture requirement crops. Climate change with expected rise of 2–4°C in the next hundred years will bring a still greater challenge for the rain fed agriculturists. Intensities and frequencies of events such as droughts and floods will increase.

Watershed Development and Management Approach

During the ancient period, village boundaries were decided upon on the watershed basis by the expert farmers in the villages. Such boundaries were socially acceptable to all the members of the system. Such age-old village boundaries are fixed at the common point of the drainage system in between the two villages. Watershed based planning of resource management has generated a wide appreciation in India, particularly for assured dividends. The concept of maintaining an ecological balance embedded in the watershed programme has also started getting attention in different sections of the society.

As the entire process of agricultural development depends on the status of water resources, the watershed with its distinct hydrological boundary is considered ideal for taking on a developmental programme. Planning and designing soil and water conservation structures such as bunds, waterways, overflow hydraulic structures, water harvesting tanks, etc., are carried out considering the expected rate and amount of runoff and flood volumes. This helps in reducing soil and nutrient loss, top fertile soil removal, improved *in situ* soil moisture and ultimately to improve crop productivity. Watershed development is fundamentally focused on conservation, regeneration and the judicious use of all the resources; both natural (land, water, plants, animals) as well as human components within the watershed area (Shinde 2014). Watershed management seeks to bring about the best possible balance in the environment between natural resources and man/ animals (Mani 2005). Since it is the man who is chiefly responsible for the degradation of the environment thus regeneration and conservation can only be possible by promoting, awakening and ensuring the participation of the people who inhabit the watershed vicinities. Watershed management is defined as the integrated use, regulation and development of the water and land resources of a watershed in order to accomplish the sustainable use of land, aqua and flora. The emphasis is on soil and water conservation on the watershed basis. Integrated watershed management involves working on the natural and human resources in a watershed in accordance with the social, political, economic and institutional factors that operate within the watershed (Hufschmidt 1991).

Principles of Watershed Management

The main principles of watershed management under Mahnot and Singh 1993, are:

i) Utilizing the land according to its capability.
ii) Maintaining adequate vegetative cover on the soil for controlling soil erosion, mainly during the rainy season.
iii) Conserving the maximum possible rainwater at the place where it falls, on arable land by contour farming.
iv) Draining out the excess water with a safe velocity to avoid soil erosion and storing it in ponds for future use.
v) Preventing erosion in gullies and increasing ground water recharge by putting in nullah bunds and gully plugs at suitable intervals.

Multiple Use Concept in Watershed Management

A multiple use perspective is required to achieve sustained and integrated watershed management, particularly in those areas, where a large rural population depends upon a variety of resources produced in upland watersheds. It may be noted that much of the intensive farming, grazing, and timber harvesting that take places in most of the areas is leading to watershed degradation, loss of bio-diversity and adverse downstream impacts.

Watershed inhabitants in many areas practice multiple use, which involves the production of goods that they require such as food, fiber, fuel and fodder. Most of the development activities are closely associated to the development and use of water resources. Thus, multiple use is being practiced on various watersheds, but whether multiple use is being properly managed for upland and downstream inhabitants is a matter of concern.

The main aim of multiple use management is to manage a natural resource mixture for the most beneficial combination at both present and future uses. It is not necessary that every watershed is managed for all possible natural resource products simultaneously. Instead, most of the watersheds are utilized for various natural resource products depending on levels of supplies and demand. Multiple uses can be accomplished by one or more of the following options (Brooks et al. 1997):

i) Concurrent and continuous use of several natural resource products obtainable on a particular watershed requiring the production of several goods and services from the same area.
ii) Alternating or rotating the use of various natural resource products on a watershed.
iii) Geographic separation of uses or use combinations, so that multiple use is accomplished across a mosaic of land management units on a watershed, with any use to which it is most suited.

It is important that effective multiple use management should accommodate to the extent possible the full spectrum of today's requirements while providing for tomorrow's needs.

Types of multiple use management

There are two types of multiple use management viz. resource oriented and area oriented. Resource oriented multiple use management represents the alternative uses of one or more natural resources. For instance, timber can be managed for lumber, fuel wood and pulp. Such management depends upon the knowledge of interrelationships-showing how the management of one natural resource affects other uses of the same resources or how one use of a natural resource affects other uses of the same resource. Resource oriented multiple use management needs thorough the understanding of the production capacities of natural resources.

Area oriented multiple use management represents the production of a mix of products and amenities from a particular area. It is important that area-oriented multiple use should consider the physical, biological, economic and social factors related to resource product development in a given area. Area oriented multiple use gets the information required to describe resource potentials from resource oriented multiple uses and then relates this to the dynamics of local, regional and national demands. Brooks et al. (1997) show multiple resources from the land area that results in several products (Table 1). It may be noted that area oriented multiple use management is not necessarily intended to replace other forms of land management but to complement them.

Table 1. Multiple resources from land area resulting several products.

Resource	Products
Water	Irrigation, Municipal or Industrial, Recreation
Timber	Lumber, Pulp, Wood, Fuel wood, Recreation
Forage	Livestock, Wildlife, Recreation
Wildlife	Consumption, Recreation
Minerals	Depends on the Type of Mineral

Integrated Watershed Management Approach

Integrated watershed management approach focuses on the assimilation of various technologies within the natural boundaries of a drainage area for optimum development of land, water, and vegetation to meet the fundamental needs of people and animals in a sustainable manner (Wani and Garg 2009). It orients to enhance the standard of living of the common people (Wani and Garg 2009) by increasing his earning capacity by offering all facilities required for optimum production (Singh 2000). In order to accomplish various objectives of integrated watershed management, various strategies are worked out simultaneously like land and water conservation practices, water harvesting in ponds and recharging of groundwater for increasing water resources potential and stress on crop diversification, use of an improved variety of seeds, integrated nutrient management and integrated pest management practices, etc. (Wani and Garg 2009).

Soil and Rainwater Management Practices

Soil and water conservation measures are aimed at management of rainwater, soil and vegetation resources in a manner that perceptible changes with regard to water resources development take place in the watershed so as to increase land productivity on a sustainable basis (Arora 2006). Not only should the surface water storage increase as a result of soil water conservation interventions, but increased ground water recharge should take place. Some of the effective and feasible soil and water conservation practices either indigenously followed or adopted through technological interventions in watershed programmes by the farmers of the Shivalik foothills in north-western Himalayas includes field bunding, pre-monsoon ploughing, terracing, contour trenching, earthing-up in maize, straw and soil mulching and tillage management (Arora et al. 2006, Arora and Hadda 2003).

Socio-Economic Development

The watershed development programme in agricultural and forest catchment's aims in soil and water conservation result in several ecological benefits viz. reduction in soil loss, development of vegetative cover, fodder production, increase in crop yields, wasteland development, etc. This in turn results in the economic development of resource poor rural communities in the region, as indicated through increased availability of fuel, fodder and commercial grass, employment generation and economic analysis.

Productivity and income generation

Watershed management programmes will not be self-sustainable, if improvement in productivity and generation of additional income does not commensurate with investment. Increased biomass and fodder production resulting from integrated management of watershed helps to change the composition of livestock to more economical animals and reduced seasonal migration of herds due to assured fodder supply during the year. The harvested rainwater in small storage tanks/ structures/farm ponds can be effectively utilized for supplemental irrigation during lean periods to boost crop production. Water harvesting structures proved to be economically viable, environmentally sound and socially acceptable (Samra 2002).

Bio-Industrial Watershed Management Approach

The term bio-industrial connotes two meanings. Firstly, bio highlights the human-centered development that the project promotes. Through agricultural inputs and social interventions, the watershed community remains at the center of the program. As their on-farm efficiency and profitability increases, their social standing is expected to do the same. Secondly, the word industrial points toward the enhancement of livelihoods and the development of a more diversified economy in the village. Beyond the promotion of on-farm livelihoods, need to garner off-farm and non-farm livelihoods for the sake of enhancing income security (McGhghy 2012).

Bio-industrial Watershed Management is watershed management plus processing industries for value addition of agricultural products before marketing them. This is the way to make the presently profitless farming in India, to be profitable (Bali 2005). Special bio-industrial zones need to be marked and infrastructure developed. First, soil and water conservation measures have to be applied. After that, the small watershed unit must be provided with assured agricultural knowledge and inputs availability. Bio-processing industries, owned preferably by the farmers, have to be developed. Natural resources are then developed by developing land, water and vegetation.

A financial system of loans and subsidies must support each bio-industrial watershed. The tenants' rights have to be secured. Bio-industrial Extension personnel must be available to do the *running about* for each bio-industrial watershed. Farming is supported financially by all enlightened governments. Farm prices are kept low to reduce poverty. But this works unfairly for the farmers. That is why most governments subsidize farming to the tune of 30 or 40 per cent. In France, subsidies touch 80 per cent. Subsidized farming keeps food prices low and helps eradicate poverty. There are three major components of the bio-industrial watershed management, i.e., resource conservation, sustainable biomass production and processing of produce.

Resource conservation in bio-industrial watershed

Soil and water conservation, within bio-industrial watersheds, is vital for rain fed agriculture. Every drop of rain has to be conserved. Soils are poor and shallow. Land is sloping and water runs off quickly. Land is to be protected with contour bunds which will level the land over time. Water harvesting has to be developed. The ephemeral streams around provide an opportunity for farm ponds which can give life-saving water. It is here that the watershed management would meet its toughest challenge. It is here that the need is the greatest, that small quantities of produce must be processed and converted into high value nutritive products which will bring good money to the farmers who must be a partner in the processing industry. Watershed programmes act as pivot to agricultural growth and development in rain fed areas.

Sustainable production of biomass in bio-industrial watersheds

Bio-industrial Watershed Management would be meaningful when marketable produce is available from different crops in quantity and quality. If what is produced is all consumed, there would hardly be any scope for processing and marketing for extra money in the pocket of the poor growers (Bali 2005). Present crop yields are low while the potential is high. This is a boon in a way. We shall have scope to expand the yield and cater to the food needs of the future high population. Those countries which have already achieved the peak productivity would not have such potential. But the crop yields must improve quickly.

China has only two-thirds of the area under agriculture compared to India, but her food production is double of India's. Wheat and rice constitute about three-fourth of the food grain production in the country, but the productivity of both these crops is lower than other countries and also below the world average. Other countries have two to three times the yields achieved in India. There is tremendous scope of increase of productivity and total food production in our country. The need is for water harvesting and management, including rainwater management, and further intensification of application of science, technology and inputs. All said and done, population control is a major area of attention. India cannot afford to multiply indefinitely. There is a school of thought which says health and education facilities are the best contraceptive. But there is a place for direct intervention also. India's programmes are on the right path. Only the speed needs to be enhanced so that poverty is reduced within a stated time frame.

Rural poverty comes in the way of the effective adoption of agricultural technologies available from research. The only way seems to be adoption of the Bio-industrial model of rural development in which processing is an essential step and which is bound to increase incomes and eradicate rural poverty.

Processing, value addition, storage and marketing

The foregoing efforts in the increase of productivity and market surpluses will make it possible to introduce processing of the watershed produce as an essential component of the watershed development and management programme. Strategies are needed for value addition to the products by supporting approaches such as structural mechanism, non-structural mechanisms and institutional approach (Cosgrove and Loucks 2015). Bio-industrial Watershed Management would be the ideal vehicle to take industries to agriculture and the rural people.

Bio-Industrial Watershed Opportunities in Hilly Region

There are many plus points to the mountains. Good climate and attractions for the tourists, an appropriate niche for horticulture of a great variety, rich biodiversity, medicinal and aromatic plants, animals well-adapted to the terrain, cooperative people, and a haven for future organic farming and food processing industries which can lift people from poverty to prosperity. People in and around the watershed are convinced for linkages between watershed conservation status and downstream hydrological benefits and the users to pay for the existing services, examples like the watershed protection, bio-prospecting and ecotourism (Tognetti et al. 2005). Regenerating watersheds in a holistic manner (watershed development) helps in revitalizing the ecosystem, the base of food sources and addressing biodiversity and sustainability concerns. There is plenty of potential for clean hydroelectricity, especially in hilly tracts and thanks to the Tehri Dam and other mini-hydroelectric projects. Even tiny projects can be installed on the old abandoned watermill sites and the new sites as well. Almost all the hill states of India are abound in the potential for cash crops like saffron, flowers, off season vegetables, vegetable seeds, mushrooms, honey, silk, wools (including the fine Angora rabbit wool), bamboo and other bio-products on which rural industries can be based. The need is there for holistic development on the watershed basis. If only agricultural production is pursued there will be the serious consequences of erosion and biodiversity disappearance affecting the future generations. In Morocco, the Sebou watershed is one of the most populated geographical zones and this watershed is equipped in various industries. Two hundred units are installed in the watershed and are mainly represented by oil factories, sugar factories, tanneries, paper factory, textile units, etc., using conserved water and providing livelihoods (Jaghror et al. 2013). In the Ethiopian watershed, industries gave impetus to improved watershed management adopting, different soil and water conservation practices, and rehabilitation of watershed through afforestation, community woodlots development and construction of micro and small-scale irrigation projects (Hoben 1995, Gebremedhin et al. 2003).

Agriculture alone is not paying, much less so in the hilly watersheds. There is an urgent need for agriculture plus industry to add value to the produce of plants and animals. In the words of Prof. M.S. Swaminathan, we have to integrate Ecology, Economics, Employment and Equity.

Contract Farming and Bio-Industrial Watershed Management

Contract farming comes close to the bio-industrial watershed model, if the whole watershed is taken for resource conservation, development and raw material production with distinct objectives: (a) ensuring regular supply of raw materials (b) avoiding incidences of distress sale (c) promoting cultivation of process able varieties of farm produce (d) preventing wastage of surplus farm produce and increasing its shelf life through processing, and (e) commercializing agriculture through contract farming.

Way Forward for Bio-Industrial Watersheds

There is a need to take certain initiatives in the watershed programmes in the near future to make Bio-industrial watershed a reality (Bali 2005).

Strengthening processing and value addition

Bio-industrial Watershed Management has the potential of ushering in a Bio-industrial Revolution in the Rural Areas, eradicating poverty. Processing components may, therefore, be added to all the current watershed programmes and the existing guidelines may be suitably amended to enable the change.

Department of bio-industrial watershed management

Existing Watershed Organizations in the States may be re-organized in order to establish a strong Department of Bio-industrial Watershed Management, to make Bio-industrial Rural Revolution a reality.

Bio-industrial watershed management coordination council

Bio-industrial Watershed Management would require close coordination between the Ministries of Rural Development, Agriculture and Food Processing Industries. A Bio-industrial Watershed Management Council may, therefore, be set up bringing all the concerned Ministries together.

Bio-industrial watershed research and training institute

Bio-industrial Watershed Model with its union with a number of departments and organizations and with a multiplicity of disciplines, needs a separate Bio-industrial Watershed Management Research Institute. In the meantime, existing research establishments should take up definite studies of the Bio-industrial Watershed Problems of different regions and different socio-economic conditions. However efficient the organization which is built up for demonstration and propagandas be, unless that organization is based on the solid foundation provided by research, it will merely be a house built on sand. It is hence important that we pay attention to strengthening the research and development infrastructure essential for sustainable food security in an era of climate change (Swaminathan 2010). Food security and environmental degradation are two of the main challenges facing humanity in the twenty-first century (Lal 2000). Protecting and strengthening watershed ecosystems is one of the main strategies.

Extension

Agricultural Extension should contain a wing on Bio-industrial watersheds. The Extension agencies should convey knowledge of various assistance schemes for the rural bio-industries. There should also be the link between rural entrepreneurs and the sources of processing technologies like the CSIR.

Processing vital for perishable produce of watershed villages

At present hardly 2% of fruits and vegetables are processed. In order to save the profitable horticultural produce, and increase rural incomes, this proportion must be brought up to 25% within 5 years.

Credit and Insurance

The agriculture and industry combination in the Bio-industrial watersheds need access to easy credit and also crop insurance. Farmers suffer for want of extensive insurance coverage, and commit suicide when caught in the debt trap. Special credit facilities will have to be set up to promote the system. Present facilities are neither adequate nor easily accessible to the rural people. Credit and insurance cover for the crop and stored produce is essential for the success of the Bio-industrial Watershed Management Movement.

Marketing

Marketing is crucial for the success of the Bio-indstrial Watershed Movement. Existing marketing structure may, therefore, be reviewed to make it more effective in bringing the bulk of the profits of processing to the primary growers.

Monitoring and evaluation

Monitoring of physical parameters is not enough. We must monitor whether poverty in the rural areas has been alleviated if not eradicated. Evaluation should similarly ascertain the improvement in the real incomes of the villagers, specially the deprived sections of the society.

Encouraging self help groups and NGOs

A rural individual is weak in economic power. Grouping under various systems is necessary. Whether it is the Cooperative, Corporate Body, Self Help Group or any other institution depends upon the local conditions and choice of the people. Genuine NGOs can play an important role in pushing forward the Bio-industrial Management Movement.

Reaching the unreached

The whole purpose of bringing industry to join with agriculture is to help the poor, deprived and the unreached sections of the society. Unless the profits of industrial processing of bio-produce are derived by the upstream growers also, poverty shall persist. The role of Government and the NGOs should be as facilitators to bring genuine Bio-industrial benefits to the rural people to bring about a Rural Bio-industrial Revolution, comparable to the Industrial Revolution of the Nineteenth Century, which never reached India.

Conclusions

Soil and water conservation practices are essential components of watershed development programme. If properly implemented through farmers' participatory approach, the soil and water conservation practices in agricultural catchments, shall enable the farmers to optimize their crop yields and also rehabilitate the erosion prone degraded lands. To make agriculture a profit giving venture in rain fed and hilly areas, on which young men would build their livelihood willingly, a processing industry; would have to be added to agriculture on the pattern of the Bio-industrial Watershed Management. The approach of watershed with agricultural and rural development activities should be converged into the bio-industrial watershed for synergy effect. Watershed Programmes of India will yield the desired results only when they are converted into Bio-industrial Watershed Programmes.

> Prof. *M.S. Swaminathan says*: "*I hope this concept (Bio-industrial Watershed Management) will get incorporated into the design of watersheds. Various International and National Conferences have endorsed late Prof. Bali's concept. I hope soon every watershed in our country will become a Bio-industrial watershed, in order to ensure work and income security to rural families*"

(Bali 2005)

References

Aher, P., A. Jagarlapudi and S. Gorantiwar. 2014. Quantification of morphometric characterization and prioritization for management planning in semi-arid tropics of India: A remote sensing and GIS approach. J. Hydrol. 511. 10.1016/j.jhydrol.2014.02.028.

Aher, P.D., J. Adinarayana and S.D. Gorantiwar. 2012. Use of morphological characteristics for multicriteria evaluation through fuzzy analytical hierarchy process for prioritization of watersheds. pp. 12–13639. *In*: 21st Century Watershed Technology: Improving Water Quality and the environment. Conference Proceedings, Bari, Italy. doi:10.13031/2013.41405.

Arora, S. and M.S. Hadda. 2003. Soil moisture conservation and nutrient management practices in maize-wheat cropping system in rainfed North-western tract of India. Indian J. Dryland Agric. Res. Develop. 18: 70–74.

Arora, S. 2006. Preliminary assessment of soil and water conservation status in drought prone foothill region of north-west India. J. World Assoc Soil Water Conserv. J1-5: 55–63.

Arora, Sanjay, V. Sharma, A. Kohli and V.K. Jalali. 2006. Soil and water conservation for sustaining productivity in foothills of lower shivaliks. Journal of Soil and Water Conservation, India 5: 77–82.

Bali, J.S. 2005. Bioindustrial Watershed Management, Concept and Strategies, SCSI, New Delhi, p. 97.

Brooks, K.N., P.F. Folliott, H.M. Gregersen and L.F. Bano. 1997. Hydrology and the Management of Watershed. Second Edition, Iowa State University, Ames.

Cosgrove, W.J. and D.P. Loucks. 2015. Water management: Current and future challenges and research directions, Water Resour. Res. 51: 4823–4839, doi:10.1002/2014WR016869.

Gebremedhin, B., J. Pender, J. and G. Tesfay. 2003. Community natural resource management: The case of woodlots in Northern Ethiopia. Environment and Development Economics 8: 129–148.

Hoben, A. 1995. Paradigms and politics: The cultural construction of environmental policy in Ethiopia. World Development 23: 1007–1021. http://dx.doi.org/10.1016/0305-750X(95)00019-9.

Hufschmidt, M.M. 1991. A conceptual framework for watershed management. pp. 17–31. *In*: Easter, K.W., J.A. Dixon and M.M. Hufschmidt (eds.). Watershed Resource Management: Studies for Asia and the Pacific. Singapore, Institute of Southeast Asian Studies and Honolulu, Hawaii, USA, East-West Center.

Jaghror, H., K. Houri, E.H. Saad, I. Saad, L. Zidane, A. Douira and M. Fadli. 2013. Physicochemical typology of the water in the watershed of Sebou river (Morocco). Environ. Sci. An Indian J. 8: 362–372.

Lal, R. 2000. Integrated watershed management in the global ecosystem. CRC Press, Florida, USA.

Mahnot, S.C. and P.K. Singh. 1993. Soil and Water Conservation. Inter-cooperation Coordination Office. Jaipur. p. 90.

Mani, N.D. 2005. Watershed Management, Principles, Parameters and Programmes, Dominate Publishers and Distributers, New Delhi, pp. 3–35.

McGhghy, B. 2012. The Community Managed Bio-Industrial Watershed in Karasanur. Borlaug-Ruan International Intern., MS Swaminathan Research Foundation (MSSRF), Chennai, pp. 1–21.

Nair, A.S.K. 2009. A new scientific management approach to water related natural disasters. pp. 143–154. *In*: Proceedings of Kerala environment congress, Thiruvanthapuram.

Samra, J.S. 2002. Participatory watershed management in India. J. Indian Soc. Soil Sci. 50: 345–351.

Shinde, S.D. 2014. Environmental Issues & Remedies in Watershed Development Programmes in Khatav Tahsil (Satara District).

Singh, R.V. 2000. Watershed planning and management.Yash Publishing House, Bikaner, Rajasthan, India.

Swaminathan, M.S. 2010. Safeguarding National Food Security in an Era of Climate Change. pp. 6–9. *In*: Agriculture Yearbook 2010, Agriculture Today, Connaught Place, New Delhi.

Tognetti, S.S., B. Aylward and G.F. Mendoza. 2005. Markets for Watershed Services. *In*: Anderson, M. (ed.). Encyclopaedia of Hydrological Sciences. John Wiley and Sons, UK.

Venkateswarlu, B. and C.A. RamaRao. 2010. Rainfed Agriculture: challenges of Climate Change. pp. 43–46. *In*: Agriculture Yearbook 2010, Agriculture Today, Connaught Place, New Delhi.

Wani, S.P. and K.K. Garg. 2009. Watershed management concept and principles. *In*: Best-bet Options for Integrated Watershed Management Proceedings of the Comprehensive Assessment of Watershed Programs in India, 25–27 July 2007, ICRISAT Patancheru, Andhra Pradesh, India.

Chapter 11

Land Evaluation: A General Perspective

K Karthikeyan,[1,][*] Nirmal Kumar,[1] Abrar Yousuf,[2] Balkrishna S Bhople,[2] Pushpanjali[3] and RK Naitam[1]

Introduction

Land Evaluation is the process of estimating the potential of the land for its best alternative use (Dent and Young 1981) or as the prediction of land performance when the land is used for specified purposes (Beek et al. 1997, Rossiter and Van Wambeke 1997, Herrick et al. 2016). It is the progression processes of evaluating assessing the performance of land when it is exploited for specific rationales purpose (FAO 1976). Land evaluation is therefore for implementing land use planning for both individual land users or collectively by groups of land users (Huizing et al. 1995, Herrick et al. 2016). 'Land' is a collective entity which takes into consideration of 'soil', 'topography', 'climate', and 'political density'. Therefore, it is an undoubtedly integrated geographical concept (physical and human geography). 'Reasonably stable' characteristics include variable but non-cyclic attributes that can be presented on temporal scale, in particular, the weather (Rossiter 2001). Land resource surveys and land use planning is linked with each

[1] ICAR-National Bureau of Soil Survey and Land Use Planning, Nagpur 440033, India.
[2] Regional Research Station Punjab Agricultural University, Ballowal Saunkhri, SBS Nagar, 144521, India.
[3] ICAR-Central Research Institute on Dryland Agriculture, Hyderabad, Telangana 500059.
[*] Corresponding author: mailtokarthik77@gmail.com

other through land evaluation (Deckers et al. 2004). Therefore, conducting a land evaluation incorporates assimilate various factors including soil properties, the ways in which different soils respond to different agricultural techniques, climatic variables, topography, geology, geomorphology, social and practical considerations (Peder 1986, Stomph et al. 1994, Várallyay 2011 and Herrick et al. 2016). It involves the execution and elucidation of basic surveys of climate, soils, vegetation and other aspects of land in terms of the requirements of alternative forms of land use (Verheye 2002 and Herrick et al. 2016). In order to evaluate varied land classes, the physical, economics and social background of the area must be considered (FAO 1976, Herrick et al. 2016).

The consequences of a land evaluation are based on the prediction of the potential use of land for several authentic or proposed land-use systems vis-à-vis depending on predictions of how each land part would perform, if it were used according to each of these systems (Rossiter and Van Wambeke 1997, Baniya 2008, Herrick et al. 2016). The outcomes of land evaluation, therefore, serve as a guide for organized land use decisions. The results from land evaluation serves as a foundation for decision making, by decision makers who have an influence on land use in a given region (Rossiter 2001, Ritung et al. 2007, Herrick et al. 2016).

The FAO Framework for Land Evaluation (FAO 1976) provides regulation for land suitability assessment in developing countries where data insufficiency often constrains modeling (Stomph et al. 1994). Numerous indoctrination methods are meant to match the consequences of land evaluation with the accessible means of governments, land users and other stakeholders, to attain optimal land use (Herrick et al. 2016). Two chief approaches can be distinguished: parametric systems based on a numerical correlation between crop performance and key land attributes (De la Rosa and Van Diepen 2002, Herrick et al. 2016), and categoric systems which classify the land into units with different use potentials according to the number and extent of physical limitations to crop growth (De la Rosa and Van Diepen 2002, Herrick et al. 2016). The novel trends in land evaluation are discussed below, including the increased use of crop simulation models (Van Lanen et al. 1992) as a tool for a more quantified appraisal (Herrick et al. 2016).

Land

The FAO (1995) defined Land as:

A delineable area of the earth's terrestrial surface, encompassing all attributes of the biosphere immediately above or below this surface, including those of the near surface climate, the soil and terrain forms, the surface hydrology (including shallow lakes, rivers, marshes, and swamps) near-surface sedimentary layers and the associated groundwater reserve, the plant and animal populations, patterns of anthropogenic arrangement and substantial results of anthropogenic activities (water storage or drainage structures, infrastructure, constructions, etc.).

This definition encompasses at least eight functions of land that go beyond the production of food (Landon 1984):

- It is the foundation of a variety of life sustaining support, through the production of food, fodder, fiber, fuel, timber and other biotic materials for human use, either directly or indirectly through allied services (such as animal husbandry including aquaculture, inland and coastal fisheries);
- Land is the pedestal of terrestrial bio-diversity by providing habitats and gene reserves for flora, fauna and microbes, above and underneath ground;
- Greenhouse gases are linked with land which acts as its source and sink and is a predominant factor of the global energy balance (albedo, absorption and transformation of radiated energy of the sun);
- The storage, flow and maintenance of the quality of surface and groundwater resources are regulated by land. Land helps to retain, filter, buffer and transform hazardous compounds;
- Land is a storehouse of raw materials and minerals;
- Land retains, filters, buffers and transforms hazardous compounds;
- Land provides the space for anthropogenic arrangements, industrial estates and social activities (such as sports and recreation), and connective space for transport of people, inputs and produce, and for the movement of flora (through dispersal) and fauna between natural ecosystems;
- Land preserves the facts of the cultural history of mankind, and is a source of information on past climatic conditions and past land uses.

Land Evaluation

Soil Science Society of America defines land evaluation as: "It is the process of appraisal of land performance when the land is exploited for specific reasons." It involves the implementation and elucidation of surveys and studies of landforms, soils, climate, vegetation, etc., in order to recognize and evaluate promising classes of land use in appropriation to the objectives of the evaluation. To be of significance in planning, the variety of land uses considered should be limited to those relevant within the physical, economic and social framework of the region considered, and the comparisons should incorporate economic considerations.

Land Use Planning

Land use planning (LUP) is a comprehensive progression based on the discourse amongst all stakeholders intended for cooperation and decision-making for a sustainable form of land use in rural and urban areas as well as commencing and supervising its implementation (Xiang and Clarke 2003, Herrick et al. 2016).

Role of Land Evaluation in Land Use Planning

Land evaluation is the most significant part of the land use planning. The function of land use planning is to direct decisions on land use in such a manner that the environmental resources are put to its most beneficial use, whilst conserving those resources for the future generations (Qian et al. 2018).

Objectives of Land Evaluation

The indispensable information about land resources that is obligatory for land use planning, progress and organization of these decisions before stakeholders (such as planners, user, farmer, government officials and politicians) is to recognize:

- Present land utilization?
- Is the up-gradation of management scenarios in the present land use are possible?
- What alternative uses of land is possible, acceptable and promising?
- Which of these uses are sustainable?
- What are the undesirable effects of all land uses are possible?
- What are essential inputs to diminish adverse effects?
- What are the benefits by each type of land use?

The final result is a number of clear recommendations, alternatives, and the appropriate type of land use together with their consequences.

Land Use Planning and Sustainable Development

Land use planning is anticipated to make a major contribution to the comprehension sustainable development in a comprehensive manner (Runhaar et al. 2009). It can facilitate the allocation of land to the use(s) that provide the greatest sustainable benefits (Agenda 21 par.10.5) with a focus on the fact that development remains within the carrying capacity of supporting ecosystems. The continuing worldwide mismanagement of soils, inadequate land use policies and ineffective implementation of soil management and conservation strategies, raises questions about how the communication of natural resources information to land use planners and decision makers can be enhanced and how this knowledge can be put to good use.

Qualitative to Quantitative Land Evaluation

Land evaluation process is mostly of qualitative, and based on expert opinion. The experts are typically soil surveyors and agronomists who interpret their field data and make them understandable to planners, engineers, extension officers and farmers. More recently, intensive studies of specific soil-related constraints, in particular soil fertility, available water, available oxygen, soil workability and degradation hazards (such as soil erosion and soil salinization) have all facilitated quantitative simulation of specific land use processes and opened the avenues for yield prediction. The advancement of information technology in the last two decades has enabled researchers to make rapid advancement in the analysis of interactions between land resources and land use and in quantitative land evaluation based on quantitative modeling of land use systems (Beek et al. 1997).

However, evaluating land evaluation and land use systems analysis in the broader context of land use planning, revealed a potential gap between technology-oriented land resource specialists, concerned with the present and future performance of the land resources, and human-oriented (social) scientists, concerned with the land users and their well-being (Beek et al. 1997).

Qualitative Methods (Dengiz and Usul 2018)

1. Land capability classification (Klingebiel and Montgomery 1961, Dengiz and Usul 2018, Girmay et al. 2018)
2. Land irrigability classification (Sys 1985, Dengiz and Usul 2018)

Quantitative Methods (Hack-ten Broeke et al. 2019)

1. Soil Index Rating (Storie 1978)
2. Land Capability Index
3. Actual and Potential Productivity (Requier et al. 1970)
4. Land suitability Classification (FAO 1979, Sys 1985, 1991)
5. Multivariate Regression Yield Model
6. Capability Index for Irrigation (Sys et al. 1991)

Land Capability Classification

Land capability classification is a qualitative scheme (Girmay et al. 2018) that was devised by the US Department of Agriculture, in the 1930s, as part of an erosion control program (Klingebiel and Montgomery 1961, Girmay et al. 2018). Land capability implies the potential of the land to sustain a number of predefined land uses in a built-in descending sequence of desirability: arable crops, pasture, woodland, recreation/wildlife. If the capability of land decreases, the land becomes suitable for fewer of the main land uses. Land capability is evaluated by comparing the distinctiveness of a land mapping unit with the critical limits set for each capability class. To obtain limits for the capability classes, expert knowledge was correlated with land characteristics. Sub-classes specify the kinds of limitation, whereas, capability units aggregate management recommendations according to technology and productivity levels of farming.

Land capability classification is predominantly helpful in the setting up of large farms and it augments land use planning, e.g., balances the need for agricultural land against urban development or forest land alongside agriculture or pasture development. In doing so, the land capability classification has made an important contribution to the development of land use planning and management (Beek et al. 1997).

Categories in LCC

LCC has three components—LCC Class, LCC Subclass, and LCC units—each of which is represented by a figure or symbol (Supriya et al. 2018).

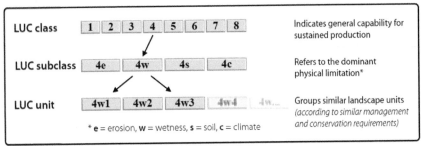

Fig. 1. Components of the LCC.

Land Capability Class

Land capability classes (Gizachew and Ndao 2008), the broadest groups, are designated by numbers 1 to 8:

Class 1 soils possess slight limitations that limit their use.

Class 2 soils possess moderate limitations that limit the choice of plants or that need moderate conservation practices.

Class 3 soils possess severe limitations that limit the choice of plants or that require special conservation practices, or both.

Class 4 soils possess very severe limitations that limit the choice of plants or that require very careful management, or both.

Class 5 soils that are subjected to slight or no erosion but have other limitations, impractical to eliminate, that limit their use chiefly to pasture, rangeland, forestland or wildlife habitat.

Class 6 soils possess severe limitations that make them usually inappropriate for cultivation and that limit their use primarily to pasture, rangeland, forestland or wildlife habitat.

Class 7 soils possess very severe limitations that make them inappropriate for cultivation and that limit their use chiefly to grazing, forestland, or wildlife habitat.

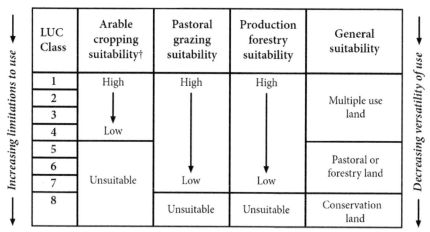

Fig. 2. The outline of LCC classes for limitations and versatility (Kingbiel and Montgomery 1961).

Class 8 soils and miscellaneous areas possess limitations that prevent commercial plant production and that limit their use to recreational purposes, wildlife habitat, watershed, or esthetic purposes.

LCC subclass

Capability subclasses are soil groups within one class. They are designated by adding a small letter to the class numeral (for example, 2e), when abbreviated forms include 'e (main hazard is the risk of erosion unless close-growing plant cover is maintained), w (water in or on the soil interferes with plant growth or cultivation), s (shallow, droughty, or stony soils), or c (extreme climatic factors, either very cold or very dry)' (Girmay et al. 2018).

In Class 1 there are no subclasses due to the fact that the soils of this class possess few limitations. Class 5 contains only the subclasses represented by w, s, or c because the soils in Class 5 are subject to little or no erosion (Girmay et al. 2018). They have other limitations that restrict their use to pasture, rangeland, forestland, wildlife habitat, or recreation (Sonter and Lawrie 2007).

LCC units

Capability units are soil groups within a subclass. The soils in a capability unit are enough alike to be suited to the same crops and pasture plants, to require similar management, and to have similar productivity (Girmay et al. 2018). Capability units are generally designated by adding an Arabic figure to the subclass symbol, for example, 2e–4 and 3e–6.

Land Irrigability Classification

The appropriateness of land for irrigation depends on physical factors such as quality and quantity of irrigation water vis-a-vis socio-economic factors such as land development costs, provision of drainage facilities, and the production costs of individual crops (McMartin 1950).

Class 1:   Lands that possess few constraints of soils, topography or drainage for sustained use under irrigation.

Class 2:   Lands that possess moderate constraints of soil, topography or drainage for sustained use under irrigation.

Class 3:   Lands that possess severe constraints of soil, topography or drainage for sustained use under irrigation.

Class 4:   Lands that are subsidiary for sustained use under irrigation because of very severe constraints of either soil topography or drainage.

Class 5:   Lands that are provisionally classed as not suitable for sustained use under irrigation.

Class 6:   Lands that are not appropriate for sustained use under irrigation.

Soil index rating

Soil Index rating is a method of rating the soil based on certain pedological characteristics that administer the lands potential utilisation and productivity capacity (Storie 1978).

Four common factors are used in determining the index rating:

- Permeability, available water capacity, and depth of the soil
- Texture of the surface soil
- Dominant slope of the soil body
- Other conditions more readily subject to management or modification by the land user (for example, flooding, salinity, alkalinity, fertility, acidity, erosion, micro-relief, etc.)

For some soils, more than one of these X conditions is used in determining the rating.

$$SIR = A \times B \times C \times Z$$

Where the factors are decimal equivalents of percentage ratings.

A = general characteristics of soil profile

B = texture of the surface soil

C = slope of the land, and

Z = other factors (reaction of surface soil, fertility, erosion)

On comparing various methods for the land suitability evaluation of wheat in the Northeast of Iran including simple limitation, number and severity of limits and parametric methods (Storie and Square root methods), it was found that the climatic characteristics of the region were suitable for wheat plantation on the all methods (Ashraf et al. 2010). Likewise, land suitability evaluation for tea in sloping lands of Guilan province in Iran was determined by using simple limitation method, the limitation method regarding number and intensity and the parametric methods including the Square root and the Storie methods for land suitability evaluation (Foshtomi et al. 2011). Results of the first and second methods showed similar marginally suitability classes (S3). According to these methods, the most important limiting factors were climate, topography and physical soil characteristics. Moreover, the results of the Storie method showed unsuitable conditions for tea cultivation (N2), except for one land unit, which had non-suitable but correctable conditions (N1).

Ashraf et al. (2010) carried out land suitability evaluation for the growth of wheat in the Northeast of Iran and they compared various methods including simple limitation, number and severity of limits and parametric methods (Storie and Square root methods), and found that the climatic characteristics of the region were suitable for wheat plantation on the all methods. Foshtomi et al. (2011) determined land suitability evaluation for tea in the sloping lands of the Guilan province in Iran. They used the simple limitation method, the limitation method regarding number and intensity and the parametric methods including the Square root and the Storie methods for their land suitability evaluation.

A rating of 100 percent expresses the most favorable, or ideal, conditions for general crop production (Storie 1978). Lower percentage ratings are allocated for fewer favorable conditions or characteristics. Factor ratings (in percentages) are chosen from tables prepared from field data and production. Certain properties are allocated a series of values to allow for variations in the properties to plant growth, development and production. Certain properties assign a range of values to allow variations in the properties that affect the suitability of the soil for general agricultural purposes (Storie 1978).

If a map unit consists primarily of 1 named soil series (a consociation), the index rating for the named soil component equals the index rating for the map unit (Storie 1978). If a map unit consists of more than 1 named factor (a complex), ratings are allocated to each named factor (soil series or miscellaneous area, such as "Rock outcrop"). Inclusions of other soils or minor components not named in the map unit name, are not used in the calculations (Storie 1978).

Map units are assigned grades according to their suitability for general intensive agriculture as shown by their Storie index ratings (Storie 1978). As per Storie (1978), the 6 grades and their range in index ratings are:

Grade 1—80 to 100

Grade 2—60 to 79

Grade 3—40 to 59

Grade 4—20 to 39

Grade 5—10 to 19

Grade 6—less than 10

Grade 1: These soils are appropriate to intensively cultivated crops that are well adapted to climate of the region.

Grade 2: These soils are good agricultural soils, although they are not as desirable as compared to Grade 1 because of a less permeable subsoil, deep cemented layers (e.g., duripans), a gravelly or moderately fine textured surface layer, moderate or strong slopes, restricted drainage, low available water capacity, lower soil fertility, or a slight or moderate hazard of flooding.

Grade 3: These soils are only fairly well suited for agriculture because of moderate soil depth; moderate to steep slopes, restricted permeability in the subsoil; a clayey, sandy, or gravelly surface layer; somewhat restricted drainage; acidity; low fertility; or a hazard of flooding.

Grade 4: These soils are poorly suited for agriculture. They are not as desirable as compared Grade 3 because of certain limitations (shallower depth; steeper slopes; poorer drainage; a less permeable subsoil; a gravelly, sandy, or clayey surface layer; channeled or hummocky micro-relief; acidity).

Grade 5: These soils are very poorly suited for agriculture and are seldom put to use. They are more commonly used as pastures, rangelands, or woodlands.

Grade 6: These soils and miscellaneous areas are not suitable for agricultural activities because of very severe or extreme limitations. They are appropriate for limited use as rangelands, protective habitats, woodlands, or watersheds.

Land Capability Index

The land capability is estimated by calculating a capability index or soil index, being the product of ratings attributed to the 6 soil characteristics:

$$Cs = U \times \frac{V}{100} \times \frac{W}{100} \times \frac{X}{100} \times \frac{Y}{100} \times \frac{Z}{100}$$

where,

Cs is the capability or soil index

U, V, W, X, Y and Z are ratings for profile development, soil texture, soil depth, colour/drainage characteristics, pH/base saturation and development of the A soil horizon, respectively.

Six land capability classes characterize the capability of the land unit for the production of the three groups exacting, moderately exacting and less exacting crops. For each group, a reference crop was used to study the relation between the capability index and yield.

Land capability index for irrigation

The various land characteristics that influence the soil suitability for irrigation are rated and a capability index for irrigation (Ci) (Albaji et al. 2014, Bagherzadeh and Paymard 2015) is calculated according to the formula:

$$Ci = T \times \frac{U}{100} \times \frac{V}{100} \times \frac{W}{100} \times \frac{X}{100} \times \frac{Y}{100} \times \frac{Z}{100}$$

where

Ci is the capability index for irrigation

T, U, V, W, X, Y and Z are the rating of soil texture, soil depth, calcium carbonate status, gypsum status, salinity/alkalinity status, drainage characteristics and slope.

The classes II to V can have the following subclasses with regard to the nature of the limiting factors:

s—constraints due to physical soil properties (T, U, V, W)

n—constraints due to salinity/alkalinity (X)

w—wetness constraints (Y)

t—topographic constraints (Z)

Land Suitability Classification

In contrast to the land capability classification systems that assess the potential of the land for general land uses, the FAO panel for land evaluation suggested a suitability classification of the land C for a specific, well described, land utilization type (Dengiz and Usul 2018).

Land suitability is the fitness of a given piece of land for a well specified land use (Dengiz and Usul 2018, Girmay et al. 2018). It is an expression of how well a land unit matches the requirements of the land utilization type.

It consists of 4 chief stages (Girmay et al. 2018):

1. Determination of the land use requirements and the corresponding land attributes
2. Characterization or quantification of the land attributes
3. Comparison of the land use with the land, and
4. Determination of the real and potential land suitability class

Classifying land suitability

The overall performance of the land when exploited for specific purposes is usually expressed in terms of suitability or productivity (Dengiz and Usul 2018). The FAO proposes a land suitability classification presented in different categories: orders, classes, subclasses and units (Girmay et al. 2018).

Suitability orders

The FAO distinguishes two suitability orders: S (suitable) and N (unsuitable).

Suitable land is land on which the sustained use for the defined purpose in the defined manner is expected to yield benefits that will justify the proposed inputs without unacceptable risk to land resources on the site or in its adjacent areas (Girmay et al. 2018).

Unsuitable land is land having characteristics which appear to preclude its sustained use for the defined purpose in the defined manner or which would create production, upkeep and/or conservation problems requiring a level of inputs unacceptable at the time of the interpretation (Girmay et al. 2018).

As such, land may be classified as unsuitable for a given use for a number of reasons:

- Proposed use is technically impossible,
- Use would cause severe environmental degradation, or
- The value of the expected benefits does not justify the expected costs of the required inputs.

The order should always be quoted in the classification symbol even when only 1 order of the land is represented in the study area.

Suitability classes

The framework at its origin permits complete freedom in determining the number of classes within each order. However, it has been recommended to use only 3 classes within order S and 2 classes within order N.

The class is indicated by an Arabic number in the sequence of decreasing suitability within the order and therefore reflects degrees of suitability within the orders:

S1 = suitable

S2 = moderately suitable

S3 = marginally suitable

N1 = actually unsuitable, potentially suitable

N2 = actually and potentially unsuitable

Although no consistent criteria are given for defining the classes, the boundaries between the suitability classes S1, S2, and S3 are generally defined by the relation between the necessary efforts and the accomplished output. As such, they change with a changing socio-economic perspective (Ziadat 2007, Girmay et al. 2018).

Distinction between suitability classes N1 and N2 of the unsuitable order is mainly based on the recognition of serious physical limitations that can or cannot be improved by major improvements such as terracing or drainage (AbdelRahman et al. 2016, Girmay et al. 2018). Table 1 gives an outline of the guidelines that can be followed when defining the suitability classes for each of the land properties in land evaluations for agriculture (Ziadat 2007). The yield percentages dividing the suitability classes vary according to economic conditions. As such a yield reduction to 40% of the optimum might be acceptable to a subsistence farmer but not to a competitive commercial endeavor (Ziadat 2007, AbdelRahman et al. 2016, Girmay et al. 2018).

Table 1. Guidelines to define the suitability classes (Sys 1983).

Suitability class		Impact on	
		productivity	required inputs
S1	high	> 80%	-
S2	moderate	40 – 80%	practical + viable
S3	marginal	20–40%	practical + restricted viability
N1	not actually, nut potentially suitable	< 20%	major land improvements
N2	actually & potentially unsuitable	< 20%	not practical nor viable

Suitability subclasses

In land evaluation reports and maps, a lower-case letter with mnemonic significance follows the suitability class symbol reflecting the kinds of limitations or the main kinds of improvement measures required (Ziadat 2007, Girmay et al 2018).

In order to keep the number of subclasses to a minimum, it is advised to report only the most dominant constraint or constraints (if two constraints are evenly severe). The lower-case letters are used to symbolize these constraints (Ziadat 2007) and have been summarized in Table 2.

Table 2. Symbolization of land suitability subclasses.

Subclass symbol	Interpretation
C	Climatic constraints
T	Topographic constraints
W	Wetness constraints (drainage, flooding)
S	Physical soil constraints (influencing the soil/water relationship and management)
F	Soil fertility constraints not readily to be corrected
N	Salinity and/or alkalinity constraints

Source: (Ziadat 2007, AbdelRahman et al. 2016)

Suitability units

This grouping is employed to identify land units possessing minor variations in management requirements. This helps to specify the relative significance of land improvement works. It is suggested to specify the land suitability units by Arabic numerals enclosed in brackets (Ziadat 2007, Dengiz and Usul 2018).

Before a land suitability class can be given to the land units, crop-specific land use requirements need to be defined and compared with the available land characteristics (Ziadat 2007).

Square root method

In this method, a quantitative scaling is assigned to each characteristic of the lands. If a specification is quite good for the intended crop, the maximum rate of 100 is assigned to it. If the same specification meets some limitations, a lesser rate is assigned to it. The square root method can be used to acquire the land and climate index as

$$I = Rmin \sqrt{\frac{A}{100} \times \frac{B}{100} \times \cdots}$$

where,

I = Land and Climate Index

A, B, ... = remaining ratings land characteristics

Rmin is the minimum rank.

Multiple regressions

The multiple regression model is often referred as a 'black box' model. Lack of data is the main limiting factor for its applicability in land evaluation as the targeted variables, which are going to be predicted and are by definition very difficult to measure. Due to the nature of linear relationship in the parameters, regression models may not provide accurate predictions in some complex situations such as non-linear data and extreme values data. As regression models need to fulfill the regression assumptions and multiple co-linearity between independent and dependent variables (Molazem et al. 2002, Zaefizadah et al. 2011).

Multi Criteria Decision Analysis (MCDA)

MCDA provides an ample collection of techniques and procedures for frame working decision problems, and designing, evaluating and prioritizing alternative decisions. Problems that are multi-dimensional in nature can be tackled with this approach very efficiently. MCDA is used to combine qualitative and quantitative criteria and to specify the degree and nature of the relationships between those criteria in order to support spatial decision making. The main purpose of the MCDA techniques is to investigate a number of alternatives in the light of multiple criteria and conflicting objectives (Voogd 1983). In order to carry out that, it is necessary to generate compromise alternatives and a ranking of alternatives according to their degree of attractiveness (Janssen and Rietved 1990).

Out of the many approaches with the MCDA spirit, the additive weighting and related procedures are the most popular (Eastman et al. 1995, Silva and Blanco 2003, Ayalew et al. 2005, Maddahi et al. 2016, Kihoro et al. 2017, Karthikeyan et al. 2018). In this additive weighting method, the decision maker assigns weights of 'relative importance' to each criterion for arriving an overall site suitability ranking by multiplying the weighted scaled value corresponding to each criterion and summing the weighted products for all considered criteria.

Weighted Linear Combination (WLC)

The WLC is a simple additive weighting based on the concept of a weighted average (Eastman 2006). WLC is the most frequently used technique, for studying spatial multi attribute decision making. This is mainly based on the concept of weighted average. In this method, the decision maker directly assigns weights of 'relative importance' to each attribute map layer. A total score is then obtained for each alternative by multiplying the importance weight assigned for each attribute by the scaled value given to the alternative on that attribute, and summing the products over all attributes. When the overall scores are computed for each alternative, the alternative with maximum overall score is selected as the best alternative. Overall suitability is calculated from the sum of the weighted data layers representing factors in the model:

$$S = \sum w_i x_i X \prod c_j$$

where, S—is the composite suitability score, xi—is factor scores, wi—is weights assigned to each factor, cj—constraints (Boolean factor), Σ—sum of weighted factors and $\prod$—product of constraints (1—suitable, 0—unsuitable).

Analytical Hierarchy Process (AHP)

AHP helps avoiding a direct assignment of weights to the identified criteria or scores to the alternatives and checks the chance of biasness. AHP typically follows a three step procedure. Firstly, a multi-level hierarchical structure of objectives, criteria, and alternatives is made. Secondly, a matrix is developed by using pairwise comparisons between the identified criteria. The comparison matrix will be a square matrix Am x m, where m is the number of criteria (m) considered for decision. Each entry ajk of the matrix A represents the importance of the jth criterion relative to the kth criterion. The entries ajk and akj satisfy the following constraint:

ajk· akj =1.

Importance of one criterion over another one in the pair is identified qualitatively between 1 and 9 based on the AHP preference scale (Table 3). The scale 1 indicates the equal importance, while 9 indicates that one factor is absolutely more important than other. The reciprocals of 1 to 9 (1/1 and 1/9) show that one is less important than other. The intensity of importance between two criteria in the matrix is filled based on experts' experience. Once the matrix A is built, it is possible to derive the weights by identifying the normalized principal eigenvector of the matrix (Saaty 1980, 2001, Ramanathan 2001).

The components of the eigenvector **w** sum to unity. Thus, a vector of weights is obtained, which reflects the relative importance of the various criteria from the matrix of paired comparisons.

The third step is to check the consistency of the judgment matrix. In AHP, an index of consistency, known as the consistency ratio (CR), is used to indicate the probability that the matrix judgment was randomly generated and is not biased (Saaty 1987).

CR = CI/RI

Table 3. Preference Scale (Source: Saaty 2008).

AHP Scale of Importance for comparison pair	Numeric Rating	Reciprocal
Extremely Importance	9	1/9
Very strong to extremely	8	1/8
Very strong importance	7	1/7
Strongly to very strong	6	1/6
Strong Importance	5	1/5
Moderately to strong	4	1/4
Moderate importance	3	1/3
Equally to Moderately	2	1/2
Equal importance	1	1

Table 4. Random inconsistency indices (RI) for N = 10 (Source: Saaty 2008).

N	1	2	3	4	5	6	7	8	9	10
RI	0.00	0.00	0.58	0.9	1.12	1.24	1.32	1.41	1.46	1.49

where RI is the average of the resulting consistency index depending on the order of the matrix given by Saaty (1980) (Table 4) and CI is the consistency index and can be expressed as

$$CI = (\lambda_{max} - m)/(m-1)$$

where λ_{max} is the largest or principal eigenvalue of the matrix and can be easily calculated from the matrix as the average of the elements of the vector whose j^{th} element is the ratio of the jth element of the vector $\mathbf{A \cdot w}$ to the corresponding element of the vector $\mathbf{w}$. m is the order of the matrix.

When the matrix has a complete consistency, CI = 0. The bigger CI means, worse consistency the matrix had (Saaty 1980, 1987). When CR was less than 0.10, the matrix had a reasonable consistency. Otherwise the matrix should be changed. The calculated results of weight would be accepted when the consistency ratio was satisfactory (Saaty 1980). In this AHP we can avoid all kind of biasness and identify the variability effectively.

Fuzzy set theory

Zadeh (1965) for the first time defined fuzzy set theory in order to quantitative defining and determination of some classes that are expressed vaguely such as very important and so on. With this model in land evaluation, mainly bell shape functions, such as sigmoid, Cauchy and kandel functions were used. It's a known fact that many elements of land properties have uncertainties. Uncertainty is inherent in decision making process, which involves data and model uncertainty. These may be from measurement errors, due to inherent variability, to instability, to conceptual ambiguity or to simple ignorance of important factors. Fuzzy sets theory is a mathematical method used to characterize and propagate uncertainty and imprecision in data and functional relationships. The fuzzy sets methodology application in land evaluation is based on the assumption that the changes in soil properties and suitability classes of land units are not crisp but gradually changing within space. Fuzzy sets are especially useful when insufficient data exist to characterize uncertainty using standard statistical measure like mean, standard deviation and distribution type (Kurtener et al. 2004). The use of fuzzy technique is helpful for land suitability evaluation, especially in applications in which subtle differences in soil quality are of major interest (Braimoh et al. 2004) and suggested that the use of krigging exploits spatial variability maps and in estimating uncertainties associated with predicted land suitability indices.

Artificial Neural Networks (ANN)

Neural networks are a computational method of data analysis that are an extension of traditional statistical methods such as regressions (White 1999), and function approximation (Hertz et al. 1991). ANN forms an "internal weight" representation of the data as to minimize an error criterion (usually least squares) without too much a priori judgment about the functional form for the data (McCleland et al. 1986). Once an ANN has been frame worked it can be trained on the existing sample data to make predictions for unknown cases. Once trained, the network is used to predict on input data. If there are many more hidden units than the data available, the network may not be able to generalize (extrapolate), and learning of the network may be hindered by the noise and measurement error in the data. This provides a means of automating classification of very large datasets. Since GIS systems are data intensive in the spatial domain, and different types of datasets could be used to make decisions and judgments, neural networks may find a useful role in capturing expertise, and in interpolating and extrapolating knowledge as an aid to decision making. Although the use of ANN requires some heuristic knowledge on the working, structure, training and interpretation of an ANN, the level of knowledge needed to successfully apply ANN is often much lower than would be the case for many other statistical methods. The basic element of an ANN consists of a number of inputs. ANN consists of a number of smaller processing elements (PEs), or nodes, joined together. PEs are usually organized into neuron layers: an input layer where that holds the response of the network to a given input, and one or more layers in between called hidden layers. PEs in these different layers is either partially or fully inter connected. These connections are associated with a corresponding weight, which is adjusted based on the strength of the connection. In the MLP algorithm, the propagation of data through the network begins with an input pattern stimulus at the input layer, the data then flow through and are operated by the network until an output stimulus is yielded at the output layer. Each PE or node receives the output layer. Each PE or node receives the weighted outputs (WjiXi) from the PEs in the previous layer, which are summed to produce the node input (Netj). The node input is then passed through a non-linear sigmoid function (f (Netj)) to generate the node output (Yj), which is passed to the weighted input paths of many other nodes.

$$Net_j = \sum W_{ji} X_i$$

where, Wji represent the weights between node I and node j, and Xi is the output from node i. the output from a given node j is then calculated from:

$$Y_j = f(net_j) = 1/(1 + \exp(-(net_j + b))))$$

The coefficient b called bias and W weights are estimated to minimize the deviations between the targets and the estimates.

Neural networks do not require data sets to be perfect, because they are inherently built for error minimization. Fischer and Gopal (1994) and Yeh and Li (2003) found that ANNs perform better than conventional models because they are better suited to adjust to uncertainties in spatial data.

Future Perspectives

In order to be pertinent to the potentialities and constraints of each land unit, these hanging land uses and management practices must be based on land evaluation results, in order to estimate its suitability and vulnerability. Future viewpoints reveal that agro-ecological land evaluation is the correct way to respond to the what, why, and how of moving towards sustainable rural development. Therefore it is pertinent to incorporate novel means of technological interventions (e.g., satellite images, digital elevation models), extracting maximum value from data (e.g., Internet-accessible databases and sophisticated modeling techniques), and mounting the accessibility of the end products (e.g., low-cost spatial viewers). As land use is a dynamic discourse and land evaluation is interested in detecting the changes, a future challenge will be to improve the efficiency of the maintenance and updating of the land use data sets. Furthermore, this will allow them to identify the representative areas (whether high-potential or critical problem areas), for more detailed inventories on medium- or large-scale maps. The procedures will also respond with the integration of geo-referenced databases, evaluation models, and results presentation, generating maps of land use options or alternatives, through land use and management decision support systems. One more current trend is the transformation of land evaluation which results into legislative instruments, e.g., for good agricultural practices or environmental legislation. It is imperative to accept that future changes of land use and management is possible by switching over to sustainable land use systems; reducing the present rates of land degradation (such as soil erosion, salinization, acidification, eutrophication, nutrient loss, soil and water contamination, bio-diversification loss); managing land-based greenhouse emissions and ascertaining carbon scrubbers for long term storage; providing an explicit basis for quantifying greenhouse gas emissions from agricultural production, and establishing the size of potential carbon sinks under various policy scenarios.

References

AbdelRahman, M., A.E. Natarajan and R. Hegde. 2016. Assessment of land suitability and capability by integrating remote sensing and GIS for agriculture in Chamarajanagar district, Karnataka, India. Egypt. J. Remote Sensing Space Sci. 19: 125–141.

Albaji, M., M. Golabi, S.B. Nasab and M. Jahanshahi. 2014. Land suitability evaluation for surface, sprinkler and drip irrigation systems. Trans. R. Soc. S. Afr. 69. 10.1080/0035919X.2014.892038.

Ashraf, S., R. Munokyan, B. Normohammadan and A. Babaei. 2010. Qualitative Land suitability evaluation for growth of wheat in Northeast of Iran. Research Jour. of Biol. Sci. 5: 548–552.

Ayalew, L., H. Yamagishi, H. Marui and T. Kanno. 2005 Landslide in Sado Island of Japan: Part II. GIS-based susceptibility mapping with comparison of results from two methods and verifications. EngGeol. 81: 432–445.

Bagherzadeh, A. and P. Paymard. 2015. Assessment of land capability for different irrigation systems by parametric and fuzzy approaches in the Mashhad Plain, northeast Iran. Soil Water Res. 10: 90–98.

Baniya, N. 2008. Land suitability evaluation using GIS for vegetable crops in Kathmandu Valley/ Nepal, der Humboldt-Universität, Berlin, 259 pp.

Beek, K.J., A. De Bie and P. Driessen. 1997. Land evaluation for sustainable land management. ITC, Enschede, The Netherlands.

Braimoh, A.K., P.L.G. Vlek and A. Stein. 2004. Land evaluation for maize based on fuzzy set and interpolation. J. Environ. Manage. 33: 226–238.

Van Diepen, C.A., G.J. Reinds and G.H.J. Koning. 1992. A comparison of qualitative and quantitative physical land evaluations, using an assessment of the potential for sugar-beet growth in the European Community. Soil Use Manage. 8: 80–89.

De la Rosa, D. and C.A. Van Diepen. 2002. Qualitative and Quantitative Land Evaluation in 1.5. Land Use and Land Cover, in Encyclopedia of Life Support System (EOLSSUNESCO), Eolss Publishers, Oxford, UK.

Deckers, J., O. Spaargaren and S. Dondeyne. 2004. Soil survey as a basis for land evaluation. *In*: W.H. Verheye (ed.). Land Use, Land Cover and Soil Sciences Encyclopedia of Life Support systems. Abu Dhabi, UAE.

Dengiz, O. and M. Usul. 2018. Multi-criteria approach with linear combination technique and analytical hierarchy process in land evaluation studies. Eurasian J. S. Sci. 7: 20–29. DOI: 10.18393/ejss.328531.

Eastman, J.R., W. Jin, P.A.K. Kyem and J. Toledano. 1995. Raster procedures for multi-criteria/multi-objective decisions. Photogram. Eng. Rem. Sen. 61: 539–547.

Eastman, J.R. 2006. Idrisi Andes Guide to GIS and Image Processing, Clark Labs, Clark University, 950 Main Street. Worcester. MA, USA.

F.A.O. 1976. Framework for Land Evaluation. FAO Soils Bulletin No 32, Food and Agriculture Organizations of the United Nations-Rome, Italy.

FAO. 1979. Soil survey investigations for irri- gation. Soils Bull. 42. Rome.

FAO. 1995. Forest resources assessment 1990. Global Synthesis. FAO, Rome.

Fischer, M.M. and S. Gopal. 1994. Artificial neural networks: a new approach to modeling interregional telecommunication flows. J. Regional Sci. 34: 503–527.

Foshtomi, M., M. Norouzin, M. Rezaei, M. Akef and A. Akbarzadeh. 2011. Qualitative and Economic land suitability evaluation for Tea (*Camellia sinensis* L.) in sloping area of Guilan. Iran. J. Biol. Environ. Sci. 5: 135–146.

Ghazanchaii, R. and A. Fariabi. 2014. Evaluation of qualitative, quantitative and economical land suitability for major crops in dezful region. Int. J. Soil Sci. 9: 120–132.

Girmay, G., W. Sebnie and Y. Reda. 2018. Land capability classification and suitability assessment for selected crops in Gateno watershed, Ethiopia. Cogent Food Agric. 4: 1 DOI: 10.1080/23311932.2018.1532863.

Gizachew, A.A. and M. Ndao. 2008. Land evaluation in the Enderta District-Tigray-Ethiopia. 28th Course Professional Master, Geomatics and Natural Resources Evaluation, 5 November 2007–27 June 2008. IAO, Florence, Italy. [Google Scholar].

Hack-tenBroeke, M.J.D., H.M. Mulder, R.P. Bartholomeus, J.C. van Dam, G. Holshof, I.E. Hoving, D.J.J. Walvoort, M. Heinen, J.G. Kroes, P.J.T. van Bakel, I. Supit, A.J.W. de Wit and R. Ruijtenberg. 2019. Quantitative land evaluation implemented in Dutch water management. Geoderma. 338: 536–545.

Herrick, J.E., O. Arnalds, B.T. Bestelmeyer, S. Brignezu, G. Han, M.V. Johnson, Y. Lu, L. Montanarella, W. Pengue and G. Toth. 2016. Summary for policymakers: Unlocking the sustainable potential of land resources. Evaluation systems, strategies and tools. United Nations Environment Programs (UNEP). Job Number: DRI/2002/PA. 27 pp.

Hertz, J.A., A.S. Krogh and A. Palmer. 1991. Introduction to the Theory of Neural Computation: Addison-Wesley Pub. Co., Redwood City, CA.

Huizing, H., A. Farshad and C. Debie. 1995. Land evaluation. ITC lecture notes. ITC, Enschede, The Netherlands.

Janssen, R. and P. Rietveld. 1990. Multicriteria analysis and geographical information systems: an application to agricultural land use in the Netherlands. pp. 129–139. *In*: Scholten, H.J. and J.C.H. Stillwell (eds.). Geographical Information Systems for Urban and Regional Planning.

Karthikeyan, K., D. Vasu, P. Tiwary, A.M. Cunliffe, P. Chandran, M. Sankar and S.K. Singh. 2018. Comparison of methods for evaluating the suitability of Vertisols for *Gossypiumhirsutum* (Bt cotton) in two contrasting agro-ecological regions, Archives of Agronomy and Soil Science. DOI:10.1080/03650340.2018.1542131.

Kihoro, J., N.J. Bosco and H. Murage. 2017. Suitability analysis for rice growing sites using a multicriteria evaluation and GIS approach in great Mwea region, Kenya. Springer Plus 2013 2:265.doi:10.1186/2193-1801-2-265.

Klingebiel, A.A. and P.H. Montgomery. 1961. Land capability classification. Handbook No. 210 Soil conservation service, USDA.

Kurtener, D., S.E. Krueger and I. Dubitskaia. 2004. Quality estimation of data collection. pp. 9101–9109. *In*: UDMS, UDMS press. Giorggia-Venice.

Landon, J.R. 1984. Booker tropical soil manual: a handbook for soil survey and agricultural land evaluation in the tropics and subtropics. New York: Longman. xiv, 450 pp. S599.9.T76 B72 1984.

Maddahi, Z., A. Jalalian, M.K. Zarkesh and N. Honarjo. 2016. Land suitability analysis for rice cultivation usingagis-based fuzzy multi-criteria decision making approach: Central Part of Amol District, Iran. Soil & Water Res. doi: 10.17221/1/2016-SWR.

McCleland, J.L., D.E. Rumelhart and the PDP Research Group. 1986. Parallel Distributed Processing: Explorations in the Microstructure of Cognition. The MIT Press. Cambridge, MA.

McMartin, W. 1950. The economics of land classification for irrigation. J. Farm Econ. 32: 553–570. Retrieved from http://www.jstor.org/stable/1233181.

Molazem, D., M. Valizadeh and M. Zaefizadeh. 2002. North West of gentic diversity of wheat. J. Agri. Sci. 20: 353–431.

Peder, A., R.G. Margaret, S. Herald and B. Wang. 1986. A Decision-Making Tool in the Developing Countries, Nowegian Computing Center, Box 335, Blindern, N-0314 Oslo 3, Norway.

Qian, L., Y. Yang, J. Xiaoqian and G. Yuntao. 2018. Multifactor-based environmental risk assessment for sustainable land-use planning in Shenzhen, China. Sci. Total Environ. 657: 1051–1063.

Ramanathan, R. 2001. A Note on the use of the analytic hierarchy process for environmental impact assessment. Journal of Environmental Management 63: 27–35.

Riquier, J., D.L. Bramao and J.P. Comet. 1970. A new system of appraisal in terms of actual and potential productivity. FAO Soil Resources, Development and Conservation Services, Land and Water Development Division, FAO, Rome, Italy.

Ritung, S., W., F. Agus and H. Hidayat. 2007. Land Suitability Evaluation with a Case Map of Aceh Barat District, Indonesian Soil Research Institute and World Agroforestry Centre, Bogor, Indonesia.

Rossiter. D.G. and A.R. Wambake. 1997. Automated Land Evaluation System ALES Version 4.65 User's Manual. Cornell University, Department of Soil, Crop and Atmospheric Sciences SCAS Teaching Series No. T 93–26, Ithaca, NY 1997.

Rossiter, D.G. 2001. Land Evaluation Processes Draft Copy. Cornell University College of Agriculture and Life Science, Department of Soil, Crop and Atmospheric Sciences.

Runhaar, H., P.P.J. Driessen and L. Soer. 2009. Sustainable Urban Development and the Challenge of Policy Integration: An Assessment of Planning Tools for Integrating Spatial and Environmental Planning in the Netherlands. Environment and Planning B: Planning and Design 36(3): 417–431. https://doi.org/10.1068/b34052.

Saaty, T.L. 2008. Decision making with the analytic hierarchy process. Int. J. ServSci. 1: 83–98.

Sayed, A.S.A. 2006. Pedological studies on soils of Esh-Milahah depression, Red sea region, Egypt. M.Sc. Thesis, Faculty of Agriculture, Al-Azhar University, Cairo, 1–190 pp.

Silva, A.C. and J.L. Blanco. 2003. Delineation of suitable areas for crops using a multi criteria evaluation approach and land use/cover mapping: a case study in Central Mexico. Agricultural Systems 77: 117–136.

Sonter R.O. and J.W. Lawrie. 2007. Soils and rural land capability. *In*: Charman, P.E.V. and B.W. Murphy (eds.). Soils: Their Properties nd Management. 3rd edition. Oxford University Press, Melbourne.

Stomph, T.J., L.O. Fresco and H. Keulen. 1994. Land use system evaluation: Concepts and methodology. Agric. Syst. 44: 243–255.

Storie, R.E. 1978. Storie Index Soil Rating. Division of Agricultural Science, University of California. http://anrcatalog.ucdavis.edu/pdf/3203.pdf.

Supriya, K., P. Kavitha, M.V.S. Naidu and M.S. Reddy. 2018. Land capability classification of Mahanandi mandal, Kurnool district, Andhra Pradesh. J. Pharmacog. Phytochem. 7: 3429–3433.

Sys, C. 1985. Land evaluation. AlgemeenBestuurvandeOntwikkelingss, Ghent, Belgium: International Training Centre for Post-Graduate Soil Scientists. State University of Ghent.

Sys, C., E. Van Ranst and J. Debaveye, 1991. Land evaluation. Part l. Principles in land evaluation and crop production calculations. International Training Centre for Post-graduate Soil Scientists, University Ghent, 265 pp.

Sys, C., E.V. Ranst and J. Debaveye. 1991. Land Evaluation. Part II. Methods in land evaluation. Agricultural Publications. General Administration for Development Cooperation, Brussels, 273p.

Van Lanen, H.A.J., M.J.D. Hack-ten Broeke, J. Bouma and W.J.M. De Groot. 1992. A mixed qualitative/quantitative physical land evaluation methodology. Geoderma 55: 37–54.

Várallyay, G. 2011. Challenge of Sustainable Development to a Modern Land Evaluation System, Land quality and land use information in the European Union. Keszthely, Hungary, JRC 61094.

Verheye, W. 2002. Land Evaluation Systems Other than the FAO System. Encyclopedia of life support systems. Abu Dhabi, UAE.

Vink, A.P.A. 1975. Land use in advancing agriculture. Springer, Berlin, Heidelberg, New York, 392 p.

Voogd, H. 1983. Multicriteria Evaluation for Urban and Regional Planning: Pion, London: 74 p.

White, H. 1999. Learning in Artificial Neural Networks: A statistical Perspective. Neural Computation. \B1\P, 425–464. MIT Press, Cambridge, MA.

Xiang, W.N. and K.C. Clarke. 2003. The Use of Scenarios in Land-Use Planning. Environment and Planning B: Planning and Design 30: 885–909. https://doi.org/10.1068/b2945.

Yeh, A.G.O. and X. Li. 2003. Simulation of development alternatives using neural networks, cellular automata, and GIS for urban planning. Photogramm. Eng. Remote Sens. 69: 1043–1052.

Zadeh, L.A. 1965. Fuzzy sets. Infor. and Control. 8: 338–353.

Zaefizadah, M., M. Khayatnezhad and R. Gholamin. 2011. Comparison of multiple linear regressions (MLR) and artificial neural network (ANN) in predicting the yield using its components in the hulless barley. American-Eurasian J. Agric. and Environ. Sci. 10: 60–64.

Ziadat, F.M. 2007. Land suitability classification using different sources of information: Soil maps and predicted soil attributes in Jordan. Geoderma. 140: 73–80.

Index

Printed and bound by CPI Group (UK) Ltd, Croydon, CR0 4YY

17/10/2024

01775709-0017